D0954965

Advance Praise for
Smart Power

"*Smart Power* paints a sharp picture of the historic challenges facing the utility industry, its regulators, and the nation at large. Peter Fox-Penner's urgent call for a bottoms-up solution relying on local, state, and regional cooperation and creativity presages the work now ongoing across the country. *Smart Power* is an essential read for policy makers looking for workable solutions for the next decade and beyond."

—**Charles Gray**, Executive Director of the National Association of
Regulatory Commissioners

"An absolutely terrific piece of work—remarkable scope and depth, while remaining accessible and pragmatic."

—**John Kwoka**, Finnegan Professor of Economics, Northeastern University

"In *Smart Power*, Mr. Fox-Penner provides a valuable and insightful analysis of where the U.S. electric power industry is headed and what it must do to successfully transition to a low-carbon environment. He recognizes that technology will transform a centralized, passive power system into one that is dynamic, interactive, and increasingly customer-centric. To succeed in this new framework, he persuasively argues that the industry must add value by providing least-cost energy services, including energy efficiency. In the case of the investor-owned segment of the industry this will require a fundamental restructuring of investment incentives by regulators. The public and cooperative sectors are better positioned, since their business model provides ample incentive to deliver least-cost energy solutions to their customer-owners. This book should be required reading for all industry regulators as they prepare to confront the challenges of this new paradigm."

—**Mark Crisson**, Chief Executive Officer of the American Public Power Association

"Few economist/engineers understand the electricity system as well as Peter Fox-Penner, and far fewer can explain it as lucidly. Whether or not you agree with every detail, his vision of the opportunities, risks, uncertainties, and tipping points of this vast and crucial industry is powerful and provocative."

—**Amory B. Lovins**, Chairman and Chief Scientist, Rocky Mountain Institute

"This book provides a thoughtful vision of the opportunities for the electric power industry to make use of new organizational and regulatory frameworks and new technologies so that it can successfully adapt to climate change, energy security, and economic efficiency challenges in the twenty-first century."

—**Paul Joskow**, President of the Alfred P. Sloan Foundation

"If you're serious about policies that place energy efficiency on a level playing field with new energy supplies, and energy policy generally, this book is essential reading."

—**Art Rosenfeld**, former Commissioner of the California Energy Commission

"*Smart Power* is the most advanced look at how climate policies will change our energy utilities, from power sources to operations to business models. It's a must read for anyone serious about climate solutions."

—**Joe Romm**, Editor of ClimateProgress.org and Senior Fellow at the Center for American Progress

"An excellent treatment of the critical issues facing the electricity industry."

—**Thomas R. Kuhn**, President of the Edison Electric Institute

About Island Press

Since 1984, the nonprofit Island Press has been stimulating, shaping, and communicating the ideas that are essential for solving environmental problems worldwide. With more than 800 titles in print and some 40 new releases each year, we are the nation's leading publisher on environmental issues. We identify innovative thinkers and emerging trends in the environmental field. We work with world-renowned experts and authors to develop cross-disciplinary solutions to environmental challenges.

Island Press designs and implements coordinated book publication campaigns in order to communicate our critical messages in print, in person, and online using the latest technologies, programs, and the media. Our goal: to reach targeted audiences—scientists, policymakers, environmental advocates, the media, and concerned citizens—who can and will take action to protect the plants and animals that enrich our world, the ecosystems we need to survive, the water we drink, and the air we breathe.

Island Press gratefully acknowledges the support of its work by the Agua Fund, Inc., The Margaret A. Cargill Foundation, Betsy and Jesse Fink Foundation, The William and Flora Hewlett Foundation, The Kresge Foundation, The Forrest and Frances Lattner Foundation, The Andrew W. Mellon Foundation, The Curtis and Edith Munson Foundation, The Overbrook Foundation, The David and Lucile Packard Foundation, The Summit Foundation, Trust for Architectural Easements, The Winslow Foundation, and other generous donors.

The opinions expressed in this book are those of the author(s) and do not necessarily reflect the views of our donors.

Smart Power

Smart Power

*Climate Change, the Smart Grid,
and the Future of Electric Utilities*

Peter Fox-Penner

Washington | Covelo | London

Copyright © 2010 Peter Fox-Penner

All rights reserved under International and Pan-American Copyright Conventions. No part of this book may be reproduced in any form or by any means without permission in writing from the publisher: Island Press, Suite 300, 1718 Connecticut Ave., NW, Washington, DC 20009

ISLAND PRESS is a trademark of The Center for Resource Economics.

Library of Congress Cataloging-in-Publication Data

Fox-Penner, Peter S., 1955–
 Smart power : climate change, the smart grid, and the future of electric utilities / Peter Fox-Penner.
   p. cm.
 Includes bibliographical references and index.
 ISBN 978-1-59726-705-2 (cloth : alk. paper) — ISBN 978-1-59726-706-9 (pbk. : alk. paper)
 1. Electric utilities—Deregulation—United States. 2. Electric power distribution—United States. 3. Power resources—United States. I. Title.
 HD9685.U5F6144 2010
 333.793'20973—dc22   2010002239

Typesetting by Karen Wenk

Printed on recycled, acid-free paper ♲

Manufactured in the United States of America
10 9 8 7 6 5 4 3 2 1

Contents

Smart Power

CHAPTER ONE

The First Electric Revolution

IN 1885, Muncie, Indiana, was a typical midwestern city. The rhythms of the city were set by the sun and the canter of horses pulling wagonloads in from the surrounding farms. The largest factory belonged to the Ball brothers, makers of the much beloved canning jars. By night, the city's only light came from smoky, flickering gas lamps. The countryside relied on candles and kerosene.

Over the next four decades, electricity transformed Muncie as it transformed the world. Shopkeepers found that smokeless electric lights were far better for attracting customers and less damaging to their goods. For the first time, mothers could allow their children to read alone at night, free of the fear of accidental but frequent lantern fires. The streets of Muncie were illuminated, and a system of twenty-five fire alarm boxes alerted the fire department much faster than a messenger could be sent by saddle.

Electricity, too, changed the Ball brothers' factory. Before 1900 a team of two glass blowers and three preteen boys worked by hand to make 1,000 jars a day. The electric machines that replaced these workers took eight men to run and—in the same amount of time—produced 42,000 jars. Historian David Nye writes, "In Muncie's foundries men seldom carried heavy loads, because an overhead crane with a powerful electromagnet could carry materials from one end of a plant to the other in less than two minutes. Three men operating it could do the work that previously required thirty-six strong day laborers."[1]

Insull's Industry

As Muncie and thousands of other cities electrified, one man was smiling. It was not Thomas Edison, J. P. Morgan, or any of the many other electric inventors or financiers of the era. It was Samuel Insull, the son of an English lay preacher,[2] who devised an industry structure and business model that enabled electricity to embark on an unbroken century of growth.

Insull rose from the personal staff of Thomas Edison to become CEO of one of the earliest utility holding companies, Commonwealth Edison.[3] Along the way he mastered beyond all others the technology and economics of power demand and supply, the importance of utility regulation, and the value of different business and financial structures.

Insull's visions of the industry rested on four pillars. First, it was cheaper to serve customers when their power use was aggregated via the largest possible web of interconnections—the system we now call the grid. Insull termed this the massing of consumption. The second pillar was economies of scale in production, or the industry's natural monopoly attributes. Today some of these scale effects have faded, but they were immutable in Insull's days and for decades thereafter.

When one's costs go down as supply goes up, what is the logical sales strategy? Sell more and charge less. Insull and the industry's finest marketing force sang "the gospel of consumption," urging customers to buy ever more power and giving them discounts when they did. This was pillar number three.

Finally, Insull recognized that an industry with declining costs, high capital needs, and intensive political interaction would gain stability and protection from regulation. He wrote:

> For my own part, I cannot see how we can expect to obtain from the communities in which we operate, or from the state having control over those communities, cer-

tain privileges so far as a monopoly is concerned, and at the same time contend against regulation.[4]

In league with progressives like Robert M. La Follette, Sr., who favored government control over trusts and other critical industries, a system of independent state agencies was established to oversee utilities and their rates.

Insull pursued his vision ceaselessly, acquiring and combining small power systems around the United States. The rest of investor-owned systems followed suit. A scattered collection of small power plants owned by municipal governments and individuals became an industry of huge, centralized utilities, with roughly one-third remaining in its original ownership form. Insull's vision of large supply, massed demand, increased consumption, and regulated rates reigned supreme. And without it, electrification might not have happened.

Insull, perhaps more than any other single person, changed American life. Over the span of the next four decades, nearly every urban home and shop got electric power and lights. Housewives who had spent an entire day doing the wash could now start an electric machine that finished in an hour. Factories saw productivity gains as high as one hundred times pre-electric levels. With a radio at the hearth of nearly every American household, and theaters soon to have electric sound and later air conditioning, came the birth of mass communication and the modern entertainment industry.

Electric power became fundamental to our military strength. Well before World War II began, war planners called for a massive expansion of power production. During the war years the War Production Board closely directed the building of transmission lines and new federal hydroelectric facilities, especially in the Columbia and Tennessee river valleys. Among other customers, the Tennessee Valley Authority (TVA) supplied massive quantities of power to the secret Tennessee laboratory that built *Little Boy* and *Fat Man*, the atomic bombs dropped over Hiroshima and Nagasaki in 1945. By that same year, U.S. electricity usage had increased 60% above prewar levels, introducing additional economies of scale that had not been possible during the Great Depression.[5]

In the decades following the war, electrification permeated every facet of the American economy. The maximum rating of a turbine generator has grown by a factor of 1,000 since the power age of America began.[6] The number of personal computers installed worldwide hit the one billion mark in June 2008.[7] Patients in intensive care are wired to as many as a dozen electrical devices. Warfare is increasingly electronic. Video screens are everywhere—even in elevators, where the average viewer watches them for thirty seconds. The average American home used approximately 138 kilowatt-hours a month in 1950; today the number is closer to a thousand (1 kilowatt-hour is ten hours of a 100-watt fluorescent bulb or about half a load of laundry).[8]

In 2003 the National Academy of Engineering convened a jury to recognize the most important technological developments of the century. The Academy looked out across a country with nearly ten thousand power plants, six million miles of power lines, and an inconceivable array of electric devices.[9] The Academy had little trouble choosing electrification as the preeminent engineering achievement of the twentieth century.[10]

But all things must pass, and after a century of dominance, the sun is setting on Insull's creation.

The Second Electric Revolution

Today the electric power industry faces challenges far larger than any in its history. These challenges are motivated by two worldwide policy imperatives. The first imperative is the need to adopt policies reducing the impacts of global climate change. Scientists and policymakers now largely agree that greenhouse gases (GHGs) are growing at a rate that will soon yield dangerously high concentrations in our atmosphere. To reduce the likelihood of severe damage from storms, droughts, disease, and ecosystem shifts, GHG concentrations in the atmosphere must be limited to less than 450 parts per million.[11]

The second policy imperative is the need for greater energy security. Imbalances between the supply and demand for oil, natural gas, and other fuels and key commodities can pose a threat to the economic stability and security of import-reliant countries such as the United States. Oil imports provide more than half of U.S. oil consumption and continue to grow. The U.S. trade deficit, which currently exceeds $1 trillion, is directly related to the cost of importing oil, which contributes an estimated $700 billion a year.[12] As world demand increases, suppliers such as Saudi Arabia, Iran, and Russia will continue to gain even more geopolitical leverage at an alarming rate: in 1980, the United States bought 25% of its oil from the Organization of Petroleum Exporting Countries (OPEC); by 2030 the figure will be 47%.

For countries dependent on imported oil, electric transportation constitutes an important new pathway toward greater energy security. Because U.S. electricity is made from many types of fuel, most of them from domestic sources, every auto propelled by electric power reduces the demand for imported oil. The development of lower-carbon transportation options could provide as much as $120 billion in consumer benefits by 2030.[13] With electric transport products about to take off, the power industry must prepare for a role it has never played before: bolstering our energy security by supplying power to an electrified transport fleet.

Climate change in particular poses an extraordinary challenge for the business of delivering electricity. Most policies under discussion call for U.S. greenhouse gas reductions of 80% by 2050—well within the lifespan of many power plants operating today. The latest science suggests even steeper cuts may be necessary.[14] To achieve this objective, the industry will have to make massive changes in its fuel sources and generating plants at a wholly unprecedented pace. A system of nearly one million megawatts, operating mainly on fossil fuels, will require a trillion-dollar retooling in the span of the next several decades.[15] In this massive reconstruction, the challenge is not simply one of swapping out old plants for new ones. Every change must be checked for its impact on reliability and integrated into the continuous reliability management of the entire region. In some cases, new transmission capacity will be needed, introducing a number of new questions and challenges.

The size and cost of the carbon reductions needed for a sound climate policy make greater energy efficiency an essential part of a sound climate policy. Energy efficiency is universally viewed as the best and cheapest means of reducing carbon emissions. But the power industry was designed to make and sell as much power as possible as cheaply as possible. Repurposing the industry to both sell and save electricity raises extremely difficult financial, regulatory, and managerial questions.

As the industry shifts its supply sources, builds transmission, and increases its energy efficiency efforts, the technologies at the core of its operations will shift dramatically. Over the next thirty years, the industry will adopt the so-called Smart Grid, and the architecture of the system will shift from one based exclusively on large sources and central control to one with many more smaller sources and decentralized intelligence. The Smart Grid will mark a total transformation of the industry's operating model—the first major architectural change since alternating current became the dominant system after the Chicago World's Fair in 1893.[16]

As the industry adjusts to these technology and paradigm shifts, its viability requires that we change its financial and regulatory footings. Technology, economics, and environmental considerations have rendered the foundations of Insull's business model obsolete. Thanks to the Smart Grid, the massing of consumption will give way to individual control. The industry's scale effects have changed dramatically, though not entirely. Far from the gospel of consumption, we now sing the praises of greater energy productivity and sustainability. Regulation, Insull's fourth pillar, remains in a form that no longer serves our objectives.

The new electric power industry will have to be designed with three objectives in mind—creating a decentralized control paradigm, retooling the system for low-carbon supplies, and finding a business model that promotes much more efficiency. These imperatives together will define the future of power. A system and a business model that each took more than a century to evolve must be extensively retooled in the span of a few decades. Many of the technologies and institutions needed for the job are still being designed or tested. It is like rebuilding our entire airplane fleet, along with our runways and air traffic control system, while the planes are all up in the air filled with passengers.

This book explores the future of the power sector in three parts. Part 1 begins by looking at how the industry interacts with its customers, including the overall level of sales and how the shift in the industry's operations enabled by the so-called Smart Grid could revolutionize it. These new grid technologies will transform electric pricing and create enormous regulatory challenges, all with little or no growth in overall power sales. We examine these issues in the next five chapters.

In the second part of the book, Chapters 7 to 9, we turn to the supply side of the industry and the need to decarbonize our power sources. We'll consider the costs of, and tradeoffs between, large-scale power sources such as coal plants and small-scale power sources close to customers. As one might imagine, the transmission system plays a pivotal role in this discussion.

Part 3 turns to the question of how utilities can structure themselves to respond to all of these challenges and remain viable investor-owned firms. This is an especially difficult question, as the industry must finance hundreds of billions of dollars of investment and retool its operating paradigm without much of an increase in power sales for many years to come. The book concludes by showing how both the industry's current business model and its regulatory structure must undergo a radical redesign to pursue a new economic mission: to sell least-cost energy services, not larger amounts of kilowatt-hours.

While we might hope that an industry this important will always find a way to keep the lights on, the same could be said of a global financial sector that collapsed in mid-2008 with astonishing speed and momentous repercussions. Even within the power industry, a much smaller set of challenges ignited the California electricity crisis of 2000, bringing on rolling blackouts, bankruptcies, and billions of dollars in increased electricity costs. Getting it as right as we can is important—for our climate, our economy, and our safety and national security.

CHAPTER TWO

Deregulation, Past and Prologue

IN 1990, the future of the power industry could be summarized in a single word: *deregulation*. The majority of policymakers and academic experts largely agreed that the power generators should follow in the footsteps of airlines, telephone companies, natural gas suppliers, and trucking firms and use markets rather than regulators to set prices. A new breed of energy companies, led most visibly by Enron, had made a very profitable transition from regulated to deregulated natural gas companies. They were intent on replicating their success in the much larger electric industry.[1]

Twenty years later the issues that absorb the industry—and that are the main subjects of this book—are climate change, energy efficiency, and the impacts of the Smart Grid. Whatever happened to deregulation? And what does this say about the future, when the industry grapples with enormous, unpredictable change?

As we are about to see, deregulation was oversold by its proponents and implemented abysmally by federal and state policymakers. Poor execution led to a crisis of epic proportion in California and a dismal track record in many other parts of the United States. While many of the problems with power markets have been fixed with much stronger oversight and better market designs, power markets still face thorny problems and a fair number of unhappy customers.

Of the twenty-three states that deregulated retail rates, at least eight have either suspended or scaled it back. Most of the remainder are reinstituting some form of governmental planning or oversight process. In a nationwide survey conducted in 2007, a majority of state regulators could not identify a successfully deregulated state, and about a third admitted they had serious plans to re-regulate their own. Even in England, where retail power deregulation has been most successful, the government is cautiously moving back toward greater utility oversight.

Most importantly, however, the nature and urgency of the problems facing the industry are not seen as problems that can be solved by less control over electric rates. Had the legacy of deregulation been different, policymakers might look to even greater scope for market forces. As we shall see, competition will unquestionably play a big role in the future power industry—but it will be in a form very different from Enron's vision of an electron market free-for-all.

The Industry's Tangled Structure

The economic and regulatory structure of the American power industry is a contraption only a lawyer could love. From the engineering standpoint, there are three vertical stages of production—generation, transmission, and distribution. Generators make the power in power plants, high-voltage lines transmit the power to substations in your neighborhood, and the small wires and equipment on the poles leading to your home or office are the distribution system. Electrons are created in the generator and flow through the grid into your appliances and lights. When a single company owns the entire system—from the generator to your meter—and sells you the power made in its generators, it is said to be *vertically integrated*.

The entire industry is not integrated, which gives rise to a framework in which different parts of the system are governed by different laws. Wholesale (or "bulk") power refers to power traded between a generator and a distributor or between two utilities, much like other wholesale markets. Under the Federal Power Act, a federal agency called the Federal Energy Regulatory Commission (FERC) has sole jurisdiction over pricing (rates) in the wholesale portion of the industry. This includes all high-voltage transmission (over large distances), but not lower-voltage local distribution systems. The FERC operates like a regulatory agency, with commissioners appointed by the president and confirmed by the U.S. Senate. And although the FERC can set the rates for sending power across any high-voltage line, it has almost no authority to order any kind of utility to build a line where one is needed. This authority resides with each state.

All transactions over the distribution system are regulated by state public service commissions (PSCs) under state laws, including the final retail sale of the power to each customer. Each state's laws set out the authority of its state PSC. In every state except Nebraska (which has only public power), the laws require that the PSC set regulated, cost-based rates for transporting power over the distribution system. Note, however, that transporting the power is legally distinct from selling it. Where there is traditional rate regulation, the PSC is also required to set cost-based rates for the sale of power to each customer class (e.g., residential, small commercial, large commercial, etc.), and the rates for transport and sale are bundled into a single rate.

To make matters more complicated, there are also generators owned by federal, state, or local government agencies. These are subject to much less wholesale and transmission regulation because they are believed to be unlikely to charge unfair rates and get away with it. In most states there are also distribution systems owned by government agencies and nonprofit, customer-owned cooperatives (co-ops) that distribute and sell bundled retail power. These government and co-op power sellers are seldom subject to state regulation because they, too, are viewed as unlikely to charge unfair prices.

This complex industry and regulatory structure is summarized in Figure 2-1. You can think of it as a wholesale market of power plants and the grid regulated by the FERC and retail distribution systems and integrated utilities rate-regulated or deregulated by the states. Alongside them all are publicly owned electric systems of all types that are usually not regulated by state or federal

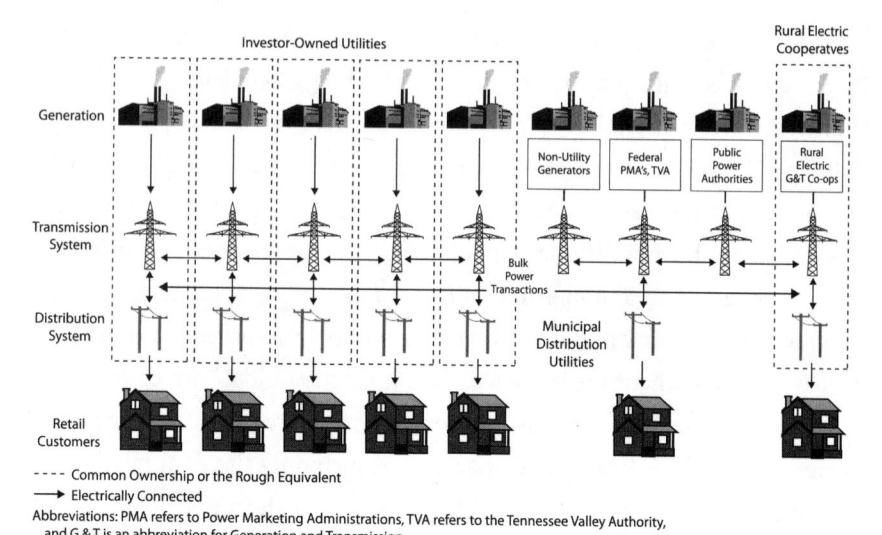

Abbreviations: PMA refers to Power Marketing Administrations, TVA refers to the Tennessee Valley Authority, and G & T is an abbreviation for Generation and Transmission

Figure 2-1. A Simple Electric System.

agencies but generally act similarly, setting rates equal to average costs that are similar to those of the IOUs but for the latter's inclusion of shareholder equity returns.

Enter Deregulation

Electric power deregulation is often thought of as a process with two giant steps, at least in theory. First, take all the power generators in the country that are now regulated and change the rules so they charge whatever prices they want. In other words, sale of the commodity electricity (kilowatt-hours), formerly purchased only from your local utility, can now be purchased from any nonderegulated generator at whatever price the market has set. However, this power can only be delivered to you via the transmission grid and the lower-voltage local distribution system. Both of these remain fully regulated. Thus, even though the market sets the wholesale prices for power itself, the rates for delivering it over the transmission and distribution wires are set by federal and state regulators, respectively. The overall price customers see on their bill is the sum of the market generation price and the two regulated delivery charges, the latter often accounting for more than half the total bill.

The second step in idealized deregulation is to make sure competition works well. Here, as in most other markets, power consumers should be able to count on antitrust officials and utility regulators to monitor and fix problems like price gouging in the new markets for generation if and when they occur.

That was the theory, at least. But due to the complex structure of the industry, and the absence of a sufficient national political consensus, restructuring did not proceed this way. Congress did not enact—nor did any president propose—a bill to deregulate generation and force all states to allow retail choice. Instead, deregulation occurred incrementally at the federal (wholesale) level and state by state at retail.

Not that people didn't try. The movement to deregulate power began with papers emanating from think tanks in the 1970s, fueled by dissatisfaction with high power rates and the cost overruns at nuclear power plants in that era. Led by Enron, a broad coalition trumpeted the benefits of electric deregulation. Retail choice was to be the unshackling of "the last great monopoly in the U.S.,"[2] a move that would lower electric rates dramatically and allow electric customers to choose the supplier of their choice. Kenneth Lay, Enron's late champion and chairman, said electric choice would be equivalent to "the largest tax cut in history."[3] Industry-backed "consumer groups" published projections of savings from deregulation as high as 43%.[4] The U.S. Department of Energy, falling in line with the prevailing winds, found that retail competition would lower rates for consumers in every state.[5] Other claimed benefits of deregulation included job creation, improvements in the reliability of service, and a cleaner environment.

Around 1994, the pro-deregulation forces began to get some traction. Although they could not convince Congress to pass a bill deregulating either the wholesale or the retail markets nationwide, they did convince the regulators at the FERC that they already had the authority to take two key deregulatory steps. First, they could create a system of "open access" in which any power generator could use anyone else's transmission system on a first-come, first-served basis to deliver power from a generator to a state-regulated distribution system. Second, the FERC started allowing some generators to make *wholesale* sales—sales only to other utilities, not actual end users—at deregulated rates. Once federal regulators enacted these key preconditions, advocates of deregulation could approach individual states. State legislatures could then vote to allow competition among deregulated retailers of power, or "retail choice," as it became known. About half the states did just this, almost all in regions where retail rates were well above the national average.

The Partial Fix

Why was deregulation introduced in this piecemeal fashion rather than as a swift, one-shot measure? Some of the reasons are purely political. Many utilities in states where electric rates were comparatively low, which included much of the South, Midwest, and West, thought regulation was working just fine in their area. If it wasn't broke, they argued, don't fix it. The argument that competition would force the cost of power down could not outweigh satisfaction with the status quo and the raw lobbying power of the IOUs in their legislatures.

The remaining reasons for easing into partial restructuring had to do with the difficulty of making sure deregulation would work. For competition to keep prices under control, there would have to be many competing generators in every area. Each of these generators has to be able to use the transmission and distribution systems to send its power on terms equal to its competitors. In other words, there are three essential preconditions to sound competition: a sufficient number of competing power generators ("deconcentration"), a transmission grid large enough to physically accommodate all competitors, and "open access" rules so that power can be shipped from generators to customers.

Another important feature of successful markets is the ability of buyers to react to price increases, that is, to use less when prices go up and more when they go down. But most of us don't even know that power prices vary hour by hour, much less how to find out what they are and then react to them. Our efforts to save power are based on our perception of annual savings, not on the ability to adjust power within a day or a week.

Unfortunately, there were very few parts of the country in which any of these conditions, let alone all, were achievable on the timetables deregulators wanted. Establishing any one of these three preconditions would be a significant political, regulatory, and financial challenge (as we shall soon see, the Smart Grid does take care of one of the three). For the most part, state and federal policymakers did not want to do the political heavy lifting required to create these conditions.

To give state policymakers the assurance that deregulation was going to work, proponents agreed to put features into state bills that were intended to protect against worst-case scenarios. These included the sorts of features one might expect: a requirement that deregulated sellers register with the state and prove their creditworthiness, periodic reports on the health of the new power markets, and so on.

The most important protection, however, was to allow any electric customers to stay with their current rate-regulated service if they did not want to switch. This option to retain regulated rates was unprecedented. It was as if airlines were to offer passengers the option of buying any ticket they chose or buying a special ticket whose rates were still set by the Civil Aeronautics Board. It was continued regulation right alongside deregulation, with customers free to choose back and forth between them. But electricity is uniquely important, and some regulators recalled that, when local phone service was deregulated, customers were furious when they were *told* by government that they had to choose a deregulated provider.

The regulated prices offered in deregulated markets became known as "provider of last resort," or POLR (pronounced "polar"), rates. One crucial decision remained: What rate should be set for POLR service? State policymakers had been told repeatedly by deregulation's cheerleaders that it was going to create *much* cheaper electricity in their states—10% at the very least, and maybe a lot more.

Although the opponents of deregulation were highly skeptical that prices would drop, they knew the proponents would argue strongly that they would. If so, there could be no harm in creating POLR rates 10% lower than current prices and frozen for five or ten years. Who would want to stay on this POLR rate if market prices were even lower? And if no one stayed on the regulated POLR prices, regulated sales would wither away, achieving the ends deregulators wanted.

The bluff worked. Governors and legislators adored this solution, as it allowed them to deliver tangible rate savings to every electric ratepayer immediately upon the enactment of deregulation. Of course, these legislated rate reductions and freezes applied only to customers who stayed with their traditional regulated utility. But this was seen as just the beginning of a new era of much lower prices for everyone.

Following deregulation, the reality was that it quickly became quite difficult for deregulated sellers to compete with the low, legislated POLR rates in many states. When the price of fuels used to make power went up, deregulated sellers had to raise their prices to cover their costs. Regulated POLR providers were either barred from raising their prices or had to wait to get permission, keeping lower prices much longer than their deregulated rivals. When this occurred, customers understandably chose to stay with the regulated POLR rates.

As this scenario played out, deregulated sellers found they could hardly ever offer small customers a price cheaper than the POLR provider, and almost no small customers switched. In many other states, however, only a few percent of residential customers switched to competitive suppliers, while roughly equal numbers switched back to regulated service. They could sometimes offer better rates to larger customers, and large customers often did migrate to competitive providers

And then came California.

The Western Energy Crisis

May 2000 began as a fairly ordinary month in California, but it did not end that way. On May 22, power prices suddenly spiked for no apparent reason. For the first time ever, prices hit a "safety valve" price cap of 75 cents/kWh—over twenty times the normal prices. Prices exceeded 25 cents (eight times the prior average) for eighty-five hours between June and September, and hit the cap another thirty-four hours in that period. Prices in all other hours continued to rise as well and with them the costs of energy purchases by power buyers throughout the western United States.

And this was only the beginning. Although prices calmed briefly in September, a November cold snap brought prices back to record levels. Supplies were even lower than in the summer, and natural gas prices peaked at approximately fifteen times their maximum level the year before. Through the first quarter of 2001, wholesale deregulated prices in the northwestern United States remained near 50 cents/kWh, the highest ever seen in this typically low-priced region.

Meanwhile, California repeatedly faced shortages of supply, threatening the reliability of the system and prompting the grid operator to declare supply emergencies on fifty-five days in 2000, and the first intentional blackouts since World War II. On December 14, 2000, the U.S. Department of Energy took the extraordinary step of issuing an emergency order requiring all generators and power marketers to sell their available surplus power to the California markets. Despite the order, supplies offered to California continued to dwindle, and during January and February 2001, the California grid operator was forced to implement its most extreme emergency procedures for thirty-two consecutive days. The system was in danger of imminent collapse and consumers experienced rolling blackouts.[6]

As you can imagine, California's governor and other state leaders were desperate to reduce power prices. Because the power markets were wholesale, and therefore under the exclusive jurisdiction of the federal government, the state could not itself impose price caps on the generators. Instead, it desperately sought to build more of its own power generators, signed contracts with new suppliers at fixed prices, and imposed some of the most successful short-term conservation efforts in history. The FERC tried a series of measures to lower prices, ultimately including caps on power prices in the entire western United States. By June 2001, these actions succeeded in taming the market, and prices fell almost as suddenly as they rose. On July 1, 2001, prices were back to about 6 cents/kWh and the crisis was over for good.

The crisis left a trail of economic devastation in California and beyond. Between June 2000 and 2001, Californians spent an estimated $33 billion more than they had paid during the prior twelve months, in addition to suffering through rolling blackouts and brownouts and the financial collapse of their utilities.[7] Power buyers in the rest of the West also paid much more for power, leading to rate increases of at least $9 billion outside of California. Largely due to the crisis, California's governor, Gray Davis, was ousted in a recall election in November 2003. But perhaps the greatest damage of all was to the idea that deregulating retail electric sales was a good idea. No elected officials wanted to take a chance that anything remotely resembling the crisis would ever occur in their state. West Virginia senator Walt Helmik summarized the views of many legislators in 2001 when he said, "Last year I thought it was a slam dunk that we were going to do this. But since then other concerns have come up, especially the situation in California."[8] Momentum toward retail deregulation froze.

The final blows came between 2006 and 2008, when the 10% POLR rate reductions enacted at the start of retail choice started expiring. In the decade since the POLR rate discounts were enacted, power plant fuel costs had risen an average of 110% and general inflation had increased prices 25%.[9] When regulators readjusted regulated POLR prices to match current average supply costs, they found that increases of 70% or more were necessary.[10] Although regulated rates (including POLR) increased by a similar amount, any illusions that deregulation was going to create enormous and visible savings were dashed for good.[11]

Deregulation's Legacy

The unhappy history of deregulation in the power industry creates an understandable air of caution among most industry policymakers. Proposals to change the structure of the industry are now met with much greater skepticism than they were in 1990. The proponents of change need to convince policymakers that consumers will be substantially better off and that nothing will go wrong—no blackouts, no messy bankruptcies, and no lessening of the quality of service.

The industry's skepticism toward deregulation is part of a larger reassessment of the idea occurring in the economics profession and policy circles. Even before the financial markets collapsed in 2008, the reappraisal was suggesting that the use of competition to achieve public ends was here to stay—but that markets are prone to very severe problems if they are not adequately designed and policed.[12] While government agencies need not set prices, they cannot sit back and assume that competition will deliver broad benefits without the careful structuring and oversight, including effective systems to protect against fraud, excessive risk taking, market power, and other problems.

Meanwhile, the era of deregulation has left the power industry with an even messier structure than we had before deregulation began. About half the country is now served by state-regulated firms who own many generators, but also buy much of their supplies, including nearly all of their renewable power. The other half of the country is served by deregulated power retailers, owned mainly by large independent generators, delivering over fully regulated wires. Utilities owned by municipalities, power districts, and other public entities are mostly unregulated and also buy some supply from the wholesale markets. Customer-owned cooperatives are similar and own many generators. Atop it all sits the wholesale power market (price decontrolled) and the high-voltage grid (price and access regulated), both overseen by the FERC.

Even without deregulation's baggage, this byzantine legal and economic structure makes sweeping organizational change in the industry quite difficult. Every change must be weighed against its impact on many different industry segments, each with different ownership, goals, strengths, and constraints. In this regard, cross-cutting changes in the power industry face adjustment cost and jurisdictional barriers similar to those bedeviling the reform of the health care and financial services sectors.

The need to rapidly reduce the industry's carbon footprint is also prompting a modest retreat toward regulation and integration. The pace at which the industry must make investments in low-carbon generation, energy-saving technologies, and new grid capacity carries with it tremendous investment risks. Regulation is designed to ensure that utilities do not earn excess profits, but also that they earn enough to keep their businesses working well. In this situation, regulation provides a modicum of insurance against some risks, such as technology failures or sudden policy shifts.[13] With climate policies forcing big changes and large, risky investments, more CEOs are thinking it worthwhile to give up the prospects of earning higher, unregulated profits in exchange for a little more protection against downside risks that aren't well understood and potentially huge.

Like every other disruptive technology, however, the changes brought on by the Smart Grid do not respect traditional jurisdictional and financial boundaries. As we shall see in the next chapter, the Smart Grid will change the entire industry's operating paradigm and open up entirely new customer relationships. Later we'll see that the need for greater energy efficiency also raises tough questions about industry structure, incentives, and responsibilities.

The future will be filled with a tension between the forces for change propelled by the Smart Grid and energy efficiency policies on one side and the perception that keeping the current structure may be more reassuring to investors, CEOs, and policymakers on the other. Deregulating more of the industry will be a challenging proposition, caught between technological change that regulators will be severely challenged to keep up with and memories of deregulatory problems they desperately want to avoid. As we will see in Part 3, the solution will be business models and reformed regulation that plug deregulated competitors into the right parts of the Smart Grid but preserve regulation and oversight in the parts of the system that still need it.

PART ONE

The Smart Grid and Electricity Sales

CHAPTER THREE

The New Paradigm

SEQUIM, WASHINGTON, was not a likely place to start the transformation of the world's electric power systems. The town—pronounced *Squim* by the locals—is known mainly as a stop on the way to hiking or kayaking on the beautifully forested Olympic Peninsula and for a climate that is ideal for growing lavender. The town of nearly six thousand hosts an annual fair proudly billed as the largest lavender event in North America.

But it was here in Sequim in early 2005 that researchers from the Pacific Northwest National Laboratory (PNNL) convinced the tiny Clallam Public Utility District—a utility too small to own a single power plant—to try something that had never been tried before. The researchers wanted to equip volunteer households with free, custom-designed computers that received electric prices set every five minutes. With the help of appliance giant Whirlpool, they would also be given thermostats, water heaters, and clothes dryers that could be programmed so that households would receive continuous feedback on the current price and quantity of power they were using and adjust their load accordingly.

PNNL's researchers knew that Clallam's power use was growing, and that Clallam's large distribution cables, known as feeders, were expensive to replace.

They were influenced by experience in wholesale power markets, where auctions are sometimes held to award capacity on oversubscribed transmission circuits to the highest bidder. In turn, wholesale markets were influenced by the work of economists such as Bill Hogan and Vernon Smith. What if prices were set to induce customers to keep their power use below the capacity of the nearly overloaded feeder? Customers who wanted to keep using power could bid for the right to use the feeder when it was filled up; other customers could bid to reduce their demand, in effect being paid by those bidding to use the feeder.

Another part of the experiment focused on the ability of the computers to help Clallam boost reliability. The computers allowed Clallam to shut down by remote control the heating element of the experiment's clothes dryers for a maximum of one minute if its operators needed a small balancing adjustment. The dryer kept spinning—only the heat cycled off and on, invisible to all but those who happened to be watching their own energy-monitoring computer at the time.

Each family was given a few hundred dollars in a bank account and told that they could keep whatever was left after their transactions were tallied at the end of the period. After a few lessons on the software, which was designed to be exceptionally user-friendly, the experiment began. Jesse Berst, editor of the fledgling *Smart Grid News*, declared it "the beginning of the GridWise era." (*GridWise* was a label the U.S. Department of Energy used for the smart grid that has since evolved into a major trade group.)

The head of the PNNL research team, scientist Rob Pratt, was amazed at the experiment's results. The 112-household marketplace successfully kept demand below the feeder's capacity at all times, though not without some fairly severe price spikes. Participating households saved an average of 10% of their power bills by managing their use and reduced their use of peak power even more. Many of the households asked to keep their equipment after the experiment ended, which unfortunately was not an option.

It was not an experiment that could be immediately replicated or scaled up. The specialized equipment cost about $1,000 per household. Customers were happy to participate because they had a guarantee that their power bill could not go up, only down—a promise reminiscent of the provider-of-last-resort rate decreases that largely undid retail choice. Price spikes were tolerated because of this guarantee, removing the need to create market power monitoring. The equipment worked seamlessly because a single government laboratory made certain it did, offering free onsite assistance when anyone had a technical glitch. Examined at close range, the GridWise Olympic Peninsula Testbed Demonstration project foreshadowed both the Smart Grid's tremendous promise and its equally large regulatory pitfalls.[1]

While the industry was busy coping with deregulation and its aftermath, power technology marched on. Communications and sensing technology became cheap and ubiquitous. Like all other technologies, nearly every kind of electrical equipment changed from analog to digital control and became progressively more sophisticated. These changes have started to unlock an entirely new vision of the power industry. To understand it, we are going to have to take a brief architectural tour of the system.

Imagine the power grid as a network of large water ponds arrayed across a vast landscape. Several narrow channels run between each pond and other adjacent ponds in every direction. The ponds are all at the same elevation. If a waterfall dumps water into one particular pond, the receiving pond naturally directs the water into all of its channels to the next adjacent ponds. They, in turn, route the water out through their other connections. Water flows freely around the network so that the level of the system is naturally even in all ponds, when there is no ability to direct the water into a specific channel.

The ponds are similar to power generators, and the channels are like the transmission system, often called the grid. Power generation is a waterfall putting water into the pond system—whichever pond the generator is attached to. The precise flow rate for water (generation) added from every generator is set by a system operator who works for the local grid.

In this pond system, using electricity means withdrawing water from a pipe that you insert into the closest pond. If you use a lot of power you need a larger pipe; if not, a straw or a piece of bamboo will do. Up to the capacity of your pipe, you can withdraw as much or as little water at a time as you want, without

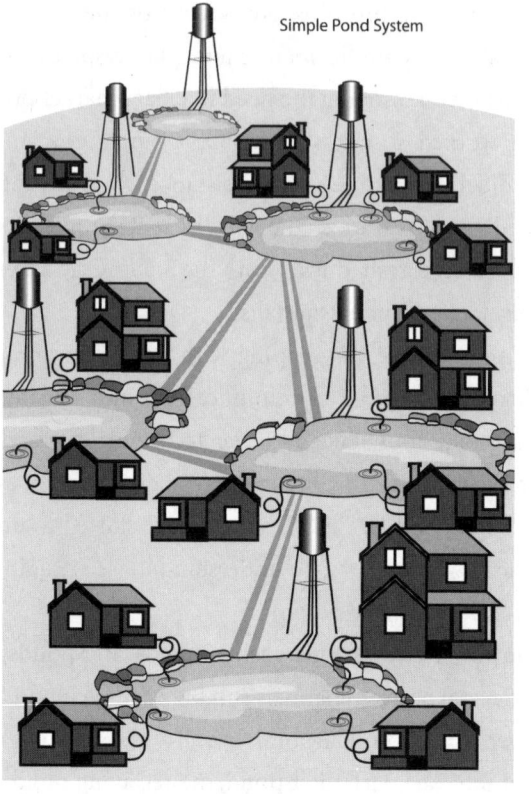

Figure 3-1. The Pond System Analogue to the Power Grid.

any sort of advance notice to the system. This is exactly like your own electric supply, where you can turn electric devices on and off at will; the only constraint is the capacity of the box of circuit breakers in your closet or basement.

The single most important aspect of power systems' architecture is the requirement for continuously perfect balance, that is, the same level of water in all ponds. All of the water that everyone is withdrawing from the ponds, the sum total of everything flowing out through the inserted pipes and straws, must equal the amount dumped into the ponds by all of the generators continuously without any interruption. This balance requirement applies on a split-second basis—the flow rates in and out must not go out of balance even for a few seconds. In the pond system, losing balance means the ponds and channels overflow; in a power system an imbalance triggers an immediate blackout. To prevent this, a power system operator who controls all of the generation in one region adjusts the power output from all of them to match total consumption on the grid on a second-by-second basis.

The requirement for perfect continuous balance may sound like a ridiculously strict requirement, but it is one that power system designers and operators conquered long ago and live with every minute of every day. In real power grids, system operators sit in secluded control centers and monitor the total power being used in each part of the country continuously. In fact, the official name for the system operators in this control center is the *balancing authority*, and the area they are required to balance is called the balancing authority area. With the help of computers that do most of the work, they adjust the generators in that balancing area to match demand, instantaneously and exactly.

The Role of the Grid

The transmission grid is the system of channels that joins all of the ponds. But why do we need these channels in the first place? Why don't we just run each pond as its own system, kept in perfect balance all by itself? Each pond could have its own system operator who simply controlled generation in the pond to match that pond's users. This would satisfy the need for immediate balance without the need to dig up the landscape and put channels all over the place.

This is the way the power industry began in the days of Muncie, Indiana. Each town had one power plant, and there were no power lines between cities or towns. Moreover, technological developments are forcing a new look at this sort of design, nowadays referred to as *microgrids*. However, with current technologies and costs, microgrids are not yet cheaper than power from the large-scale grid. In other words, if you want an electric power supply that is extremely reliable—that is, very rarely has blackouts—at the lowest possible price, you need a fleet of large generators and a grid interconnecting them.

Importantly, it is the combined desire for high reliability and lowest cost that creates the system we have. Were it not so, we would never have built the large-scale grid. The goal of the power system is to provide nearly 100% reliable service to everyone using electricity, regardless of the immediate amount of power they are using. The goal of providing reliable service is made difficult by the fact that all power generators break down unpredictably without notice, even when annual maintenance is faithfully done. When a generator "trips"— turns itself off suddenly, much like blowing the occasional fuse in the house— the requirement for immediate balance nevertheless holds.

A generator tripping off is akin to a large water tower that had been filling one of the ponds suddenly stopping. To maintain the exact water level in that pond, and in all ponds, one of two things has to occur. Either you turn on another source of water exactly as large as the one you lost instantly (a "reserve generator") or you immediately shut off downstream water users, whose total use at that moment equals the supply you've lost. Either one or a combination of these must happen to maintain immediate balance, although the consequences for the users are dramatically different.

Large power generators trip off roughly 2 to 10% of the time, aside from scheduled downtime for maintenance and refueling. There are also periods of weeks when the generators must be taken out of service for maintenance or refueling. These frequent outages don't cause blackouts because the system operators keep reserve power plants ready at all times, fully operational and ready to start instantly, much like keeping an idling car at the curb outside your house. For every 100 megawatts of plants on the system, operators keep about 15 megawatts of spare capacity; about 5 megawatts or one third of this will be idling at any time.

The grid, or interconnected system of ponds, itself allows the balancing authority to reduce the number of reserve megawatts it must maintain to achieve the same reliability by sharing these needed reserves. If one power plant goes out in any pond, a reserve unit in any other pond connected by the channels can kick in and make up the difference if the channels are wide and deep enough. Instead of each pond needing one spare generator as large as the one in service, a number of ponds can share a single reserve unit.

Numerically, this turns out to be a large savings. Take the extreme example of one single power plant supplying a completely isolated system. The one plant is either on or it has tripped off, always a possibility. To create nearly 100% reliable supply, you need a second generator just as large to provide backup or "reserves." If each generator costs $100 million, you need $200 million worth of generators total. One of those plants will sit idle nearly all the time, making power only when the first plant trips or is down for maintenance.

Roughly speaking, with a system of good channels between ponds, one extra generator can provide the same backup power or reserves to about *eight* other equal-sized generators in different ponds. The customers in each pond share the cost of the one backup generator, so they have to pay only about $12.5 million (one-eighth of $100 million) for having reserve capacity when they need it, rather than having to pay for a plant all by themselves that sits idle most of the time.

Of course, you still have to pay for the transmission lines between power plants, that is, the grid. As it happens, the transmission lines needed to provide reserves are usually much cheaper to build than the equivalent number of power plants, so that it is substantially cheaper to build fewer reserve generators and more power lines. There are several other economic advantages to interconnecting many generators via a network of lines—and, as we'll see much later, these advantages apply to decentralized minigrids as well as huge high-voltage networks.[2]

The One-Way Grid

Another extremely important feature of the power grid is its one-way nature. Among the millions of electric customers in any one region, the balancing authority really has no idea who is using what in the way of electrical devices at any one time. The only information that is communicated back to the authority in real time is the *total* amount of power being used by all customers in each portion of the balancing area—hundreds of thousands of customers or more.

Since each of us is free to turn on and off any of our electrical devices at will, and across all utility customers we own vastly different appliance collections, we are likely to use vastly different amounts of power. It would be both unfair and grossly inefficient to charge every one of us the same amount for our power use each month. The industry solved this problem by creating the traditional (nowadays often called "dumb") power meter still used today in many places.

Dumb meters cumulate the total electricity that you use over the course of a month (or normal billing period). The power company reads your meter each month and determines how much to bill you for. Since they don't know when you used the power, the price they charge per kilowatt-hour can have blocks based on cumulative monthly use or different charges in summer versus winter, but generally they don't get much more complex than this.

When the power grid was originally developed, this was really the only feasible approach. There was no Internet or Wi-Fi, of course, nor could devices be equipped with communications microprocessors and sensors. Under these conditions, there was a near-total disconnect between the industry's instantaneous balancing function and the utility's pricing and billing activities. Balancing was done by system operators based on aggregated use amounts communicated to them in real time, but which they could control only by large-scale disconnection in emergencies; monthly charges were set based on the cumulative amount used every month (quantity) and a simple price (rate) schedule. The main signaling function of the rate schedule in the distant past, and still continued in many parts of the United States, was to encourage greater consumption by progressively reducing average price as monthly use increased (so-called declining block rates).

This economic and technical architecture is what led us to today's industry structure and business model. Utilities offer a service defined by these attributes: your immediate use is controlled only by you (up to a very high limit) and can change as frequently and rapidly as you choose, using whatever devices you choose; your service will be continuous and reliable regardless of this variability, with very few blackouts; and you will be charged the average costs for the total amount of electrical energy you use over a billing period (month).

Unpacking the Monthly Power Bill

One economic feature of power systems we haven't discussed yet is the cost of making and delivering power. The most expensive ingredient in the making of power is the cost of the fuel consumed in the power plant: this is true for plants fired by natural gas and coal. The exceptions are for those renewables whose fuel is free (e.g., solar power) and nuclear power whose fuel costs are very low. The second-largest costs are the annual costs of owning and maintaining power plants, which apply to every form of generation—renewable and fossil, large and small. The rest of the costs—water use, nonfuel supplies, labor, and administration—are not very large compared to fuel and capital.

As part of their balancing duties, operators have to choose which power plant to turn on next as aggregate power use climbs over the course of a day. Some power plants are cheap to run, but they must be kept running around the clock. Other power plants are technically good at turning on quickly, which is good when more power is needed during the middle of the day. However, the plants that are most controllable are also the most expensive per kilowatt-hour made.

This means that the cost of making a single kilowatt-hour changes by a striking amount during the day, especially on very hot or cold days when electricity use is high. In the dead of night, the cheapest plants are making power for about 3 to 5 cents/kWh. As power use rises in the morning hours the next type of plant turned on costs about 6 or 7 cents. During the 100–200 hours each year when demand is the highest, making power often costs 8 to 20 cents/kWh or more, because the oldest, least efficient plants are turned on, and the cost of making them available to run these few hours during the year must be recovered.

This is illustrated in Figure 3-2. The left-hand side shows actual hourly prices during 2009. On days with mild climatic conditions and on weekends, prices don't spike up much each day, but on some days prices spike up quite a lot. On the right, the figure magnifies one day with high prices and illustrates why this happens. As demand kept rising on that day, system operators turned to increasingly expensive plants. The cost of running these more expensive plants drives up the hourly price.

A dumb electric meter adds up all of the kWh used over the course of a month regardless of when that power was made and how much it cost to make. Some homes use a lot of power during the expensive mid-day period, while others use most of their power at night. If those two homes used the same monthly total number of kWh, and they had a dumb meter, the power company has to

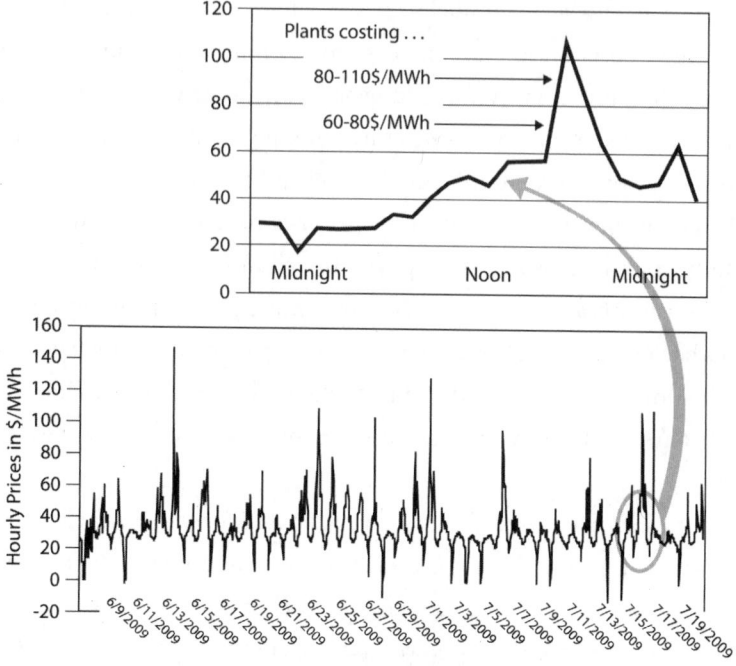

Hourly Prices in the PJM Wholesale Power Market

Figure 3-2. Using More Expensive Power Plants to Meet Peak Power Loads.
As power use goes up over the course of a day, prices go up along with demand. This occurs because system operators must turn on (dispatch) progressively higher-cost power plants to balance supply and demand instantaneously. In the one July day that is expanded, prices increase between noon and midnight because operators first dispatched plants costing $60–$80/MWh instead of the plants costing $50/MWh or less used prior to noon. Soon after, operators must dispatch plants costing $80–$120/MWh. In the PJM wholesale market, prices are based on bids rather than costs, but if competition is strong plants will bid approximately their costs.
Source: Based on data from Ventyx 2009.

charge them the same amount for monthly service because it doesn't know when each house was using power. An executive I know likens this to weighing your grocery cart when you check out at the supermarket and charging you per pound of groceries in the cart, without prices for any of the specific items you chose to buy that day, whether it be caviar or pet food.

Utilities everywhere are now starting to install *smart meters*. Among other functions, discussed in a moment, smart meters record the power used by a customer hour by hour. This allows utilities to set different prices for power sold in different parts of the day and then bill accordingly. The more power you use in the expensive parts of the day, the more you pay at the end of the month, and vice versa.

Smart meters and time-based pricing open up a panorama of new possibilities. First, setting time-based prices and billing them becomes much easier, and more pricing options can be offered. Second, smart meters can also work with smart appliances that can be programmed to automatically respond to price changes and other user commands to shift their use around within a day or even a week. This is what the residents of Sequim could do in their experiment. Finally, these meters make it easier to integrate small-scale generators and storage on customers' premises. Any meter smart enough to keep track of use by time period can also keep track of self-produced power, shift production patterns around, and even figure out when to store electricity for later use in the rare cases where storage is available.

There is no change whatsoever in the requirement that power grids always be in balance. However, smart meters and time-based pricing create a new operating paradigm. Instead of system operators manually balancing the system by turning on power plants, the users of the system can self-balance the system by adjusting their own use, and their own self-production, in response to hourly prices.

This is an elegant and intelligent new operating paradigm. Today, when electricity use rises during a heat wave, grid operators don't change anyone's power price and instead just crank up increasingly expensive and inefficient power plants. Under the Smart Grid, they will signal that prices are rising as they turn to these more expensive plants. Users who can shift their demand to later, cooler hours, will turn things off automatically via their preprogrammed smart appliances. If enough users reduce demand, system operators will stop turning on additional plants, and prices will stop going up.

At present, the size of the feedback loop that has been implemented is still fairly small and simplistic. System operators are still in charge of maintaining moment to moment balance, and while they use prices and demand response (DR) programs today as one helpful tool, they control plants manually and continuously—and will for a long time to come. Nevertheless, the combination of time-based pricing and the technological ability for customers to respond to price signals marks the beginning of a new era in the industry.

Enter the Smart Grid

If you've heard any energy speeches by candidate or President Obama or seen any of the full-page ads by the likes of Google and IBM, you've probably heard of the *smart grid*. This term has been used quite broadly in many ways, but what it really means is *combining time-based prices with the technologies that can be set by users to automatically control their use and self-production, lowering their power costs and offering other benefits such as increased reliability to the system as a whole*. It is also often more simply described as the marriage of modern information technology (IT) and the electric system. Some definitions emphasize that customers will have greater control over their energy use, others that the Smart Grid will better enable local small-scale power production, and still others that the system will be more reliable and more secure. These "definitions" simply choose to emphasize one of the many benefits of the Smart Grid over the others. For the most part, however, it all comes in one quite complicated package.

Some discussions make little distinction between smart meters and the Smart Grid as a whole. This isn't quite right, since smart meters are only one small, albeit critical, part of the new world. Another common oversimplification occurs when discussions blend the high-voltage, upstream parts of the grid with the local distribution systems. This is a little misleading, because the marriage of IT and power is different at these two levels of the industry. The paradigm shifts we have been talking about occur mainly downstream, in the local distribution companies that are either state regulated or nonprofit.

In the transmission systems (i.e., the high-voltage parts of the grid), the impact of smart technologies is quite different. In this part of the industry, the amounts of power controlled and traded are huge, handled by very large lines and system controllers who control hundreds of power plants. This part of the grid has already been using hourly pricing as well as direct control of plants and lines to balance the system and trade energy for many years. Moreover, the engineers who design and operate system controls have long used some of the most advanced computing and control tools available. "The way we look at it," one transmission engineer recently said, "the grid's been smart for a while. It takes some pretty sophisticated tools to monitor, dispatch, and control electricity flow."[3]

While this is true, computing technologies are evolving so quickly that new Smart Grid technologies are constantly faster and better at monitoring the current status of all power lines. In the future, systems will be able to forecast reliability problems minutes or hours before they occur and allow operators to implement preventive measures. They can also diagnose what went wrong when a power line trips out much faster than they could before and possibly reroute plants and lines to avoid problems. S. Massoud Amin, a professor of electrical engineering, likens the advances to flying a modern jet plane:

One of the most important of these enabling technologies is the proposal to "fly" the grid more like the way an advanced jet fighter is actually flown. Modern warplanes are now so packed with sophisticated gear as to be nearly impossible to operate by human skill alone. Instead they rely on a battery of sensors and automatic control agents that quickly gather information and act accordingly. . . . In avionics, sensing parameters like the fighter's angle of attack with respect to the position, speed, and acceleration cause automatic controllers to assist the pilot in stabilizing the aircraft via adjusting wing flaps, ailerons, or the amount of engine thrust to achieve a more optimal flight path. The grid equivalent of this would be a heightening of the "situational awareness" of the grid and allowing fast-acting changes in power production and power routing, thus altering the stream of electrical supply and demand on a moment-by-moment basis.[4]

In short, the upstream impacts of the Smart Grid will be to lessen the likelihood and severity of blackouts and to operate the system more efficiently overall. This is all to the good, but it is just an incremental improvement over the current structure and paradigm.

Downstream, for end users, the impacts of the Smart Grid are potentially profound. Customers will face electric prices that vary within each day, and they will have far more information and control over their power use and costs. With software simple enough to run on a cell phone, they'll monitor the energy used by several appliances linked to their home network, controlling them immediately or programming them to react to prices. With the touch of a button you will be able to program your air conditioner to turn off fifteen minutes out of every hour when hourly electric prices exceed a certain set-point. Yes, you'll be a little warmer, but you'll also save good money. And for the majority who don't want more complex power, appliances will all come preprogrammed so users can connect them seamlessly at factory default settings.

By its technical architecture, the Smart Grid will also encourage small-scale local generation, commonly referred to as *distributed generation* (DG), which we'll explore more deeply in Chapter 7. The widespread use of small generators will force the grid to become bidirectional and create many systemwide, but hard-to-measure, cost savings along with many regulatory challenges we'll examine shortly.

Electricity Storage

Another change brought on by the Smart Grid will be the greater use of electricity storage. The technologies that allow customers and utilities to communicate and share grid control will easily accommodate storage devices. Cheap, large-scale storage is correctly called a disruptive technology because if power can be stored cheaply the entire paradigm of immediate balance is obliterated. First off, balancing the grid gets infinitely easier—rather than worrying about turning power plants on and off constantly, you just let the batteries do the balancing. There is no need to turn power plants on in perfect synchronization with demand. Instead, just let the batteries charge when there's less demand and discharge when there is more. And if everyone can store enough power onsite for days or weeks, maintaining the grid in its current state isn't really necessary: it just needs to operate enough to recharge everyone's storage units, much as we refill our gas tanks once or twice a week. If electricity could be stored as cheaply as gasoline, the utility industry would have the same structure as the oil industry, with electric filling stations at every major crossroads.[5]

In fact, storage is not much more or less disruptive than the other technologies and functions of the Smart Grid. Distributed generation sources and customers' demand control have impacts similar to storage and face similar economic and regulatory challenges. All of these technologies are part of the same overall paradigm shift.

The New Paradigm

Customer control systems responding to prices, smaller local generators (DG), and much greater levels of storage are the three physical landmarks of the downstream Smart Grid. This new technology suite will enable customers to observe and control their electricity use as never before and to participate in keeping the grid balanced. A system that previously flowed power only from large central sources to downstream customers will flow in both directions from locally based generation and storage. New power management systems will respond automatically to hourly prices and utility signals. These features are illustrated in many pictures of the Smart Grid such as the one shown in Figure 3-3.

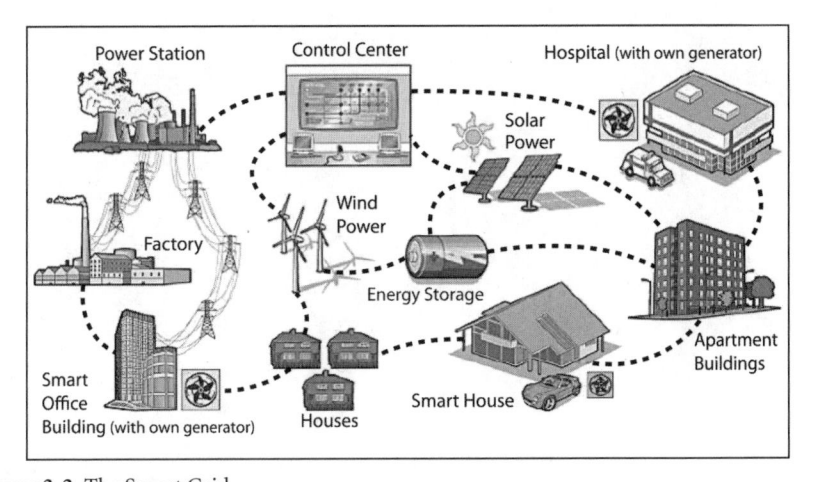

Figure 3-3. The Smart Grid.
This figure shows the Smart Grid as a system linking every element of the power system, from large generators (upper left) to homes and offices (bottom). The control center (upper middle of figure) sends out price and control signals throughout the system and keeps it perfectly balanced.
Source: Based on a graphic from *The Economist*, used by permission.

The Smart Grid will also redefine the concept of what it means to build and operate the grid. Although human system operators will still have the final say—like the fighter pilot who can still overrule the instruments, if he's fast enough—the intelligence and self-mediating role of prices will tend to self-balance the system. System operators can rely less on what was previously the only means of keeping grid balance—turning power plants on and off. They'll also have much more refined, and often self-activating, blackout-preventing tools on a scale they scarcely imagined.

As tantalizing and revolutionary as these changes will be, they will not come without adjustments and costs. Methods of setting electric rates developed over decades, and that many electric customers greatly prefer, will have to change dramatically. The magnitude of the investments involved and the breadth of potential benefits demands that Smart Grid investments be well evaluated and well implemented. It's going to be one heck of a ride.

CHAPTER FOUR

Smart Electric Pricing

I F WE CAN PUT in place the infrastructure of the Smart Grid, we will have
made a major advance on our way to a more efficient, more secure, and
climate-friendly power system. To deliver its full benefits, however, the new system must still be managed day by day and hour by hour. That means delivering
an effective set of economic signals to producers and users in the form of more
flexible and accurate electric prices.

To system operators and the hundreds of power plants they balance, the
costs of making power unquestionably change each hour (even within an hour).
There is a very old value in power system engineering that is calculated to represent this constantly changing marginal cost known as *system lambda*.

Nearly all hour-to-hour "spot market" transactions between utilities use
some form of hourly price based on system lambda or the result of an hourly
computerized auction that matches buyers and sellers. If you set your Internet
browser to the site of the New England Independent System Operator (ISO)
(http://www.isonewengland.org), one of seven centralized hourly spot markets
in the United States, you'll see an "LMP price ticker." This ticker scrolls through

that hour's price for 1 MWh power from their auction at each of fifteen delivery points across New England.

Setting the prices that *retail* customers pay equal to these hourly wholesale prices (plus a number of fixed delivery fees) is known as *real time prices*. Real-time prices send accurate signals, but they also require constant vigilance and adjustment, and they can easily change by 300% or more over the course of a day. Because many customers find this a bit too much to cope with, the industry has developed a number of less complicated time-varying rates. Because prices tend to follow a daily pattern—highest between lunch and dinner, medium during morning and late evening, and lowest in the middle of the night—one can create stair steps or "blocks" of rates that go up as you move into the peak period. These rates, known as *time-of-use*, or TOU, have been historically most common because they do not require the real-time communication of price to the customer, which until recently was cost-prohibitive. Customers only need to know what time prices go up each day, and they can read this from a "bill-stuffer" or find it on the utility's Web site.

The terminology in this area is a little obscure. Time-varying or time-differentiated rates are sometimes called *dynamic pricing*. *Demand response* (DR), is a broader concept that refers to all policies and programs that get customers to shift their use around. Sending price signals is one of the best ways to do this, but there are other, older approaches, such as allowing utilities to control customer loads directly.[1] These "direct load control" programs work well, but as the cost of price signaling keeps dropping and the Smart Grid era dawns, the use of prices to induce customers to change their load rather than controlling them directly is gaining favor.

Because the price difference for TOU blocks are averaged across the whole summer period, they are not nearly as large as the true production cost differences between mild and severe weather days. In *critical peak pricing* (CPP), the utility can be given the flexibility to set very high rates, perhaps five or ten times the usual, during just the ten to twenty days when there is a spike in demand (e.g., a heat wave). Using pagers and the Internet, customers can be warned one day in advance, so they can make plans to shift their use the next day.[2] Figure 4-1 summarizes the main approaches to time-varying rates.

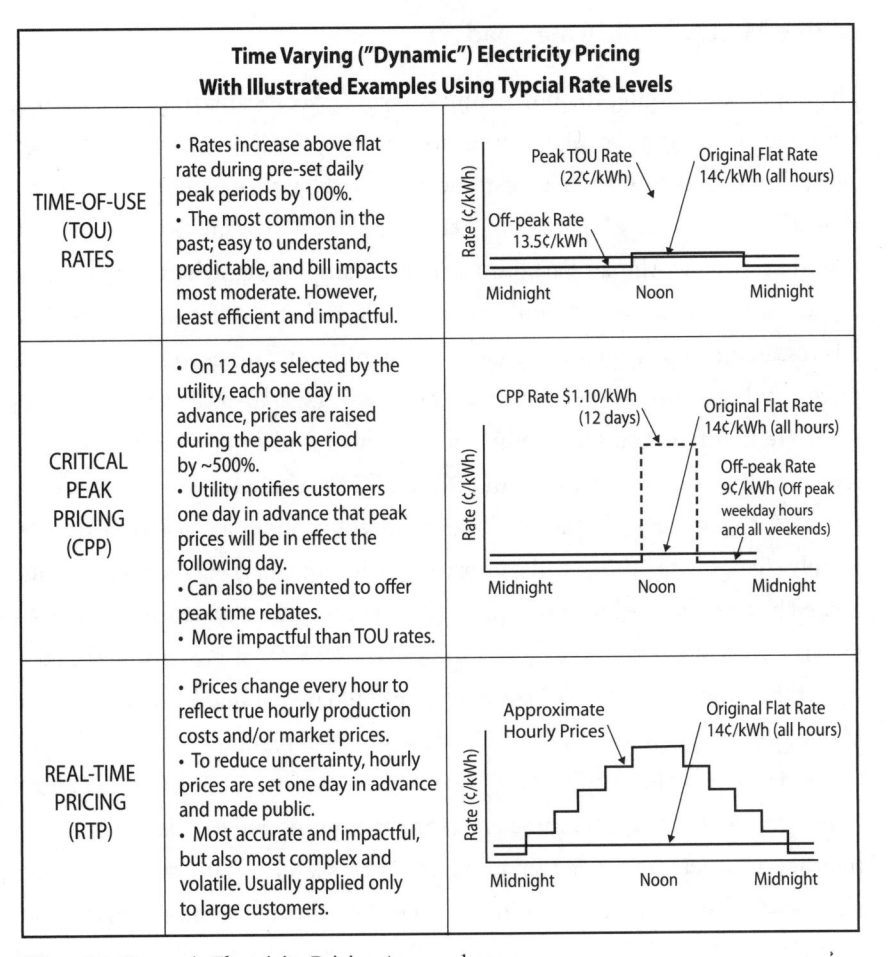

Figure 4-1. Dynamic Electricity Pricing Approaches.
Sources: Ahmad Faruqui, Ryan Hledik, and John Tsoukalis, "The Power of Dynamic Pricing," *Electricity Journal*, April 2009, and Ahmad Faruqui and Lisa Wood, "Quantifying the Benefits of Dynamic Pricing in the Mass Market," *The Brattle Group*, prepared for the Edison Electric Institute, January 2008.

Another thing to clarify about all regulated time-varying rates is that they are invariably designed to be profit neutral to the utility. In other words, the utility's short-term profits don't change when these rates are implemented—the changes in customer payments induced by these rates equal the utility's cost savings. But while utilities are neither better off nor worse off in the short run, customers benefit handsomely.

Saving Money by Shifting Load

The impact of charging customers time-varying prices is amazing. Even though many electric customers can't change the time they wash clothes or surf the net, decades of results confirm that customers find small ways to shift their use in response to dynamic rates. These small changes add up to dramatic impacts.

My colleague Ahmad Faruqui has been championing dynamic pricing for decades and has studied carefully nearly every utility that has implemented it. His research confirms that customers shift significantly in response to dynamic prices, and the response gets better as the peak versus off-peak cost differences are more faithfully translated into prices. Ahmad finds that typical TOU rates reduce peak power demand by about 5%, and CPP achieves more than twice the savings, or up to 20% of peak use.[3] When you think about it, this is remarkable. Simply changing the time profile of prices—without changing the total revenue earned by a utility one bit, and without giving customers any new technology—means that customers stop using 1 out of every 10 MW of power in peak periods. If this could be sustained nationally, we could defer building about 100,000 MW of power generators—about 200 medium-sized plants.

So far we have been talking only about how to set the profile of prices versus time. Other than to say that these prices are communicated to customers via the Internet or pagers we haven't said how customers *know* what prices are changing to or what they can do when they find out. In the fully realized Smart Grid, the prices will not only be communicated to customers, they will be conveyed immediately to preprogrammed smart devices, which will react automatically. Kurt Yeager, a former head of the Electric Power Research Institute, uses the catchy phrase *prices to devices*.

At the moment there are only a few commercially available household devices that can be programmed to take hourly price signals and adjust automatically. The most common are programmable thermostats, which change heating or cooling settings by time of day or whenever prices reach a certain point. In the future, all major appliances will have this capability. (One manufacturer, Whirlpool, has said that all its appliances will be Smart Grid–compatible by 2015).[4] Another device, known as an *in-home display*, simply receives and displays the hourly price; one popular example is an orb that glows different colors depending on each hour's price level. The buzzword for all these devices that facilitate customers' price-driven or improved manual control of their appliances is *enabling technologies*.

As would be expected, when customers have enabling technologies they shift much more of their peak power use to lower-priced periods. Under TOU rates average customers shift roughly 5 to 10% of their use. Under CPP, the amount of the shift roughly doubles, to as much as 20 to 30%. In a recent pricing pilot by Baltimore Gas and Electric, customers who adjusted their use saved 26 to 37% of their on-peak power and saved over $100 a year.[5] This level of response is more indicative of what we should expect when the Smart Grid becomes more widely deployed and significant numbers of customers have in-home networks and fully controllable appliances.

With or without enabling technologies to help them, when customers shift demand from peak to off-peak periods power systems experience a host of valuable benefits, some of them immediate and others spread over the course of many years. These benefits must be measured by regulators and divided between the utility and its customers, making regulation of these rates more challenging than simple (but horribly inefficient) flat rates. For reasons explained in a moment, state regulators control the allocation of these benefits even in states where retail choice is still in effect.

Especially in deregulated retail markets, there is an extraordinary way that the cost savings from DR are spread from those who do shift their demand to all other power buyers. In deregulated markets, the hourly spot price that everyone pays is set by the single highest-price power plant turned on by system operators to balance supply and demand. (These markets run like auctions where the highest bidder wins.) On very hot days, it isn't unusual for this plant to cost 12 cents/kWh or more. This phenomenon is illustrated in Figure 4-2, Panel 1.

When customers reduce their demand in this hot hour by shifting it to an off-peak period, the system operators don't need to turn on the 12-cent plant; load is balanced with the most expensive plant at, say, 10 cents/kWh. The market price that hour for everyone is 10 cents, not 12 cents, for all of the power purchased from the market (see Figure 4-2, Panel 2). The hourly price also

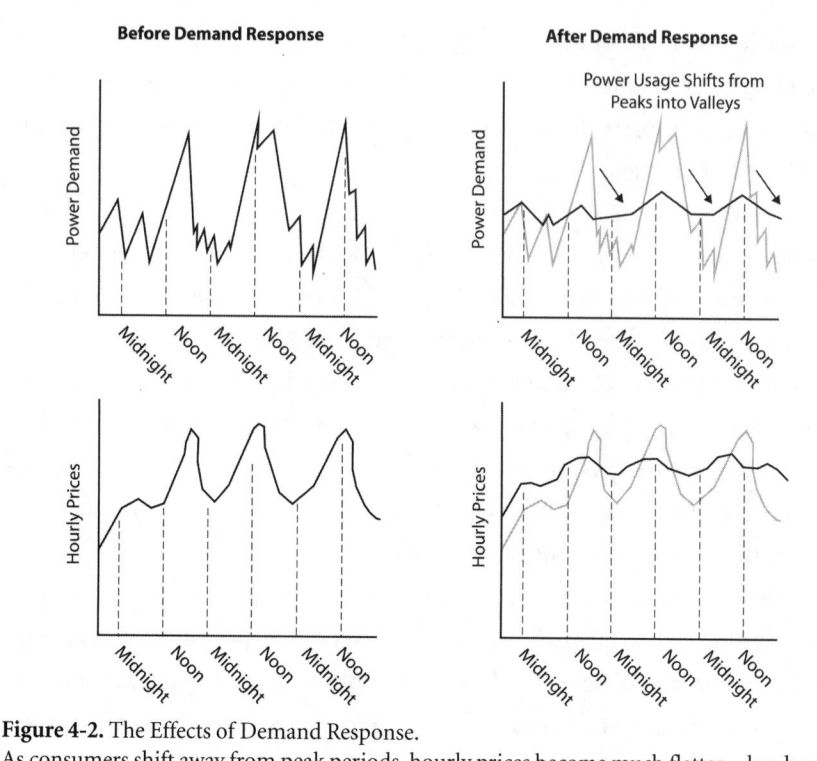

Figure 4-2. The Effects of Demand Response.
As consumers shift away from peak periods, hourly prices become much flatter—less hour-to-hour volatility. All customers benefit from lower prices to varying degrees, including those who did no shifting on their own but nonetheless pay rates lower than they would have been absent shifting.
Sources: Ahmad Faruqui, Ryan Hledik, and John Tsoukalis, "The Power of Dynamic Pricing," *Electricity Journal*, April 2009, and Ahmad Faruqui and Lisa Wood, "Quantifying the Benefits of Dynamic Pricing in the Mass Market," *The Brattle Group*, prepared for the Edison Electric Institute, January 2008.

influences many longer-term contracts, so the benefits go beyond the spot market as well.

It turns out that it only takes a small percentage of all customers shifting their use, typically about 5% of demand, to reduce prices substantially for everyone else. The amount of money saved by everyone else—who really did nothing—dwarfs the amount of money saved by the folks who shifted their load. My *Brattle* colleagues have studied this phenomenon very carefully for many utilities using models of how they dispatch power plants and the response of each type of customer to different pricing schemes. In one major study, they found that a dynamic pricing program that prompted customers to shift just 3% of their demand during the 100 highest-use hours in 2005 in the PJM market (the Pennsylvania–New Jersey–Maryland Interconnection) would save the shifting customers a midpoint estimate of $17.5 million immediately. Customers who did not shift their loads benefited from the prices reduced by the shift, saving a midpoint $120 million—over five times as much.[6]

Another benefit of DR comes from the fact that lowering the peak demand reduces the need to build power plants and the associated grid. Remember when the system operator didn't need to turn on the plant that cost 12 cents? If there is enough DR in the system, system planners figure that they'll never need the 12-cent plant. If they don't need it, customers who are shifting their loads are helping the utility avoid building a whole power plant.

This *avoided capacity cost* is often the largest benefit from DR, even larger than the energy savings to all customers.[7] In fact, DR from both traditional and Smart Grid sources is expected to eliminate 80% of all peak power growth through 2016.[8] However, unlike energy savings, avoided capacity costs don't flow to all customers immediately and are much more difficult to measure and allocate. We return to this issue, which is even more critical to small generators, shortly.[9]

Incidentally, there are other dimensions beyond hourly prices that are important for sound electricity pricing. One of them is to adopt higher per-kWh prices as one's total cumulative monthly use of power goes up, or *increasing or inclining block rates*. This is something that can be done today, even with "dumb" monthly meters. Prices that decline with higher usage levels were one of Samuel Insull's innovations, but they were designed for an era with declining long-term marginal costs.[10] Now that new power supplies cost more than existing plants it makes no sense to give customers quantity discounts, which is what declining block rates really are.

Dynamic Pricing—Barriers and Resistance

Dynamic pricing and DR are such a good idea that you might be wondering why nearly a century has gone by without these prices becoming the norm. In

BOX 4-1

Demand Response, Energy Efficiency, and Climate Change: Cousins, Not Clones

One would think that charging time-varying prices and shifting demand from peak to off-peak should be classified as a form of energy conservation, and oftentimes demand response (DR) and energy efficiency are lumped together. However, the total amount of energy and carbon saved from DR programs is much, much smaller than the dollar savings. Obviously, changing *when* you use power but keeping the amount the same is not the same as lowering the amount you use. Moreover, when prices are reduced for everyone else, their natural response is to use a little more power, not less. At the same time, it turns out that when customers see any form of time-varying prices, some of them are motivated by these prices to save electricity.

When all of these effects are added up, there is usually a small net savings induced by dynamic pricing and DR. A recent estimate of the efficiency impact of DR across all classes by my *Brattle* colleague Ryan Hledick estimates a 2.6% savings across all customer classes.* All in all, DR is a great cousin of energy efficiency policies, but it is not a substitute for the fundamental process of replacing inefficient energy-using capital, or building in more efficiency in the first place.

Roughly the same is true of CO_2 emissions from the power sector. The pure shifting of power use from peak to off-peak times doesn't save much carbon, sometimes even increasing emissions slightly in coal-heavy regions of the country. However, more aggressive DR is generally carbon-saving due to two effects. First, more aggressive DR generally brings with it more efficiency savings, hence lower emissions. Since renewable portfolio standards require that renewable plants be built, fewer fossil-fueled plants are built. Second, more aggressive DR lowers the need to build more power plants.

---

*Hledik, "How Green Is the Smart Grid?" *Electricity Journal*, April 2009.

the first place, it costs money to convert a utility's systems to dynamic pricing, and regulators always want to ensure that benefits exceed costs. Beyond this, particular resistance has come from several directions.

The new equipment utilities have to install, including smart meters and the systems that allow them to send out prices and record hourly use, are referred to as *advanced metering infrastructure* (AMI). Think of AMI as the starter version of the Smart Grid, missing most of the sophisticated customer controls I've described but enabling many of the core functionalities. For example, AMI allows dynamic pricing, but it also allows utilities to read meters electronically, eliminating the meter readers who used to visit each home each month, and to save other costs as well. Although costs continue to decline, it costs approximately $200 to $500 per customer to install an advanced infrastructure, including the communications hardware and software and other support systems.[11]

Probably the single biggest barrier to dynamic pricing has been utilities and regulators hesitating at incurring these costs. At $200 to $500 per customer, the bill adds up to billions for a large utility, adding to rates at a time of already high electricity prices and economic retrenchment. To get regulators' permission to spend this kind of money, utilities put together a business case. They add up the cost savings they expect to achieve from changes in Smart Grid–enabled operations and pass on to customers via lower rates from all aspects of their starter Smart Grid. As Smart Grid investments become better understood, utilities increasingly find that they can assemble a package of AMI outlays that provides substantial net benefits to customers over its lifetime.

Nevertheless, because the Smart Grid is such a new area and overlays are large, consumer representatives have various concerns, including the possibility that utilities will spend too much or buy a system that works poorly or is soon obsolete.[12] Others are concerned that the added data on customers' hourly use, whether in the hands of utilities or third-party vendors, will be an invasion of privacy. Still others, who see retail deregulation as a failure, distrust dynamic pricing because it has a similar ideological heritage—a view that I think is wholly understandable but misplaced.[13]

As one might expect, concern over the cost of implementing smart meters has been greatest among older and less tech-savvy customers. "We vigorously oppose the mandatory imposition of these smart meters in peoples' homes," notes Marti Doneghy, a representative of the American Association of Retired Persons (AARP). "Everybody has to pay for this change, and a lot of the 50-plus population simply isn't that interested." A more ominous note is struck by none other than Richard Thaler, coauthor of *Nudge*, who likens the Smart Grid to programming a VCR and bluntly says, "If it involves people actually doing something, it won't happen."[14] Smart Grid developers are keenly aware of this, and think they have the problem licked, but no one can say for sure.

A second source of opposition comes from the fact that, regardless of the product they are buying, customers dislike complex or volatile prices. This is why true real-time pricing, the most accurate but volatile form of dynamic pricing, is rare overall and largely confined to nonresidential customers. The other forms of dynamic pricing use approximations, but they get most of the benefits with much more predictable hourly rates.[15]

Many customers have grown to view flat rates as a "right" that is fair because everyone pays the same price. Some households and businesses can't change their use patterns, such as the lunch counter that does nearly all of its business between noon and 2 p.m. There is no technology on the horizon that fully relieves the bill increase for this customer group from dynamic prices—although energy efficiency is a good start. However, careful research has repeatedly shown that only a few percent of all customers have net higher bills on well-designed dynamic rates.[16]

Concerns over the costs of smart meters go hand-in-hand with the added complexity of setting proper dynamic prices. Unlike flat rates, which are relatively simple accounting calculations, dynamic prices require estimates of customer price responsiveness and other econometric measurements unfamiliar to nearly everyone other than economists. You need all these measurements in order to make sure that the revenues earned by the utility after dynamic prices are put in place do not lead to either excessive profits or excessive losses, both of which regulation is supposed to prevent. In addition, the peak price must reflect the capital costs avoided by the utility—a measurement challenge we're about to explore in more detail.

All sources of resistance notwithstanding, the combination of better control technologies, concerns over electric rate increases, prodding and funding from the Obama administration and the Federal Energy Regulatory Commission (FERC), and the need to tap every cost-effective source of carbon savings are overcoming a century of price structure inertia. According to the Edison Electric Institute, by 2015 almost fifty-two million customers (36% of the total) in twenty-eight states will have smart meters;[17] the Obama administration recently accelerated this with $3.4 billion in smart meter grants. And, as the current chair of the association of state regulators Fred Butler observed, "You can't have a smart grid with dumb rates."[18] With smart meters on more than a third of all customers, dynamic pricing is poised to take a quantum leap in the next decade. That, in turn, will flatten the price curve and save customers a great deal of money, along with a little carbon.[19]

Smarter electric pricing will be the single most important hallmark of the Smart Grid. Finding the political power to shift a nation built on flat power rates onto dynamic prices will be regulation's first giant adjustment to the Smart Grid. For regulators, however, this is just the opener.

CHAPTER FIVE

The Regulatory Mountain

Do you want to know what keeps me up at night? Not my programmers. Not my investors. Not my health care costs. It's state regulators.

—CEO of a leading Smart Grid firm

THE SMART GRID is a collection of technologies that enable a whole new way of operating power systems. But from the standpoint of utility CFOs and regulators, the Smart Grid is a collection of new kinds of transmission and distribution investments, each yielding unfamiliar new products and service streams. Utilities, regulators, and other stakeholders will have to evaluate these investments by measuring their value to customers, their impact on utility rates, and how customers and generators who use the new capabilities are charged for their use.

The unenviable job of balancing the many benefits of utility Smart Grid investments—fuel saving, avoided capital outlays, and complex systemic costs—with the capital costs of new systems will fall to state regulators.

Over and over, regulators will have to decide whether to put ratepayer money into large, unfamiliar investments with the promise of enormous but hard-to-measure benefits. The sort of tensions this creates are nicely highlighted in a report by the Illinois Smart Grid Initiative. The Initiative first noted that regulators ordinarily compare the cost of new utility investments to the benefits ratepayers receive in the form of lower utility costs or better service. The Initiative went on to note that

> [m]uch of the cost to deploy smart grid technologies relates to upfront capital investments—and, many of the benefits are external to the utility operations, but are predicated on this initial capital investment. Traditionally, utility investment decisions are based on achieving the lowest present value of the revenue requirement (i.e., the annual level of revenue that the regulator allows to be collected in rates). . . .
>
> . . . However, in the case of smart grid investments, much of the reduction in future expenses are realized by the consumer or the larger community, not the utility. . . . Using the traditional approach to valuing investments will almost assuredly show, at least at this time, the smart grid investments will not reduce utility expenses enough to justify investment. But if one considers the potential benefits of using a smart grid that are outside of the utility, this calculus may change.[1]

This effort will tax the capabilities of even the best regulatory agencies in the country and highlights the value of getting them more resources. But even with the best and brightest talent to draw on, the Smart Grid's regulatory challenges will be monumental and will take time and patience to wade through en route to a transformed future.

The Value of Not Building Things

One of the largest quantifiable benefits from the Smart Grid is *avoided capital costs*—the costs saved by either deferring or never building more upstream generators, transmission lines, or distribution systems. For the types of *distributed generation* (DG) that provide energy mainly at peak times, the value of the avoided capital can easily be five to ten times as large as the value of the saved energy. These benefits flow quite similarly from demand response (DR), where the customer chooses not to use more peak power, thereby avoiding the need to build a peaking plant, and distributed generators providing peak power, which substitutes for rather than avoids the upstream system. In both cases, the value of these resources depends on the capital cost of the system the utility *would have built*—the 12-cent plant in the example earlier in Chapter 4.

Measuring the costs of large and complex systems a utility *would have built* in the absence of DG and DR scattered all around its system presents challenges most regulatory practitioners love to hate. These calculations spawn acrimonious debates about the precise size, location, attributes, and cost of the large generators a utility would have built—the same debates that began in the 1980s after Congress required that each state's regulators determine avoided generation costs for each of their utilities, which said utilities would pay to certain "qualified" independent power producers. Thanks to this requirement, nearly every state has developed a method of setting avoided generation costs every few years, and there are many analytical approaches and computational models to draw on. The process is cumbersome and contentious, but it happens.[2]

You may be asking yourself whether we might spare ourselves all this regulatory debating and use deregulated markets to set avoided capacity costs. It's a great idea, but it doesn't quite work for the avoided capital costs. We are trying to measure the difference in costs between two electric systems of equal quality and reliability, one slightly smaller than the other. Because the electric power system is a system of strongly interconnected plants and lines, you can't measure the true costs of a smaller system without doing a sort of hypothetical redesign of the whole nearby grid. It's a bit like asking your architect how much cheaper your new house would be if you reduced the length of one internal room by 5 feet. The architect couldn't give you an answer until he or she redesigned the whole house to accommodate the smaller room.

Wholesale spot markets around the world are gradually adopting locational pricing, meaning that spot prices differ in every hour for many locations on the high-voltage grid. This makes prices vary not only by hour, but also by geography for each hour. This introduces a new way to value transmission, and therefore the avoided costs of not building transmission lines. One can think of the value of transmission in any hour as the difference in locational prices between two nodes on the system. For example, if prices in Philadelphia are $60 per MWh this hour, while prices in Baltimore are $75 per MWh, then the value of the transmission system between the two cities must be $15 per MWh in that hour.

Strictly speaking, this is accurate, but we build transmission lines based on their value over their lifetime of 40 years, not based on a single hour's value. However, for the purpose of paying those who use DR to curtail their load in one hour an avoided *transmission and distribution* (T&D) cost, locational price differences are a reasonable approximation. Several organized wholesale markets use approaches like this, and while the idea is conceptually sound, proper implementation is quite complex. To give a flavor for the complexity, here are the market rule requirements, simplified and stripped of their jargon, that two experts recently noted were necessary:

1. The system operator's ability to obtain good forecasts of the savings that will occur in every location on the grid as prices rise in every hour
2. A sloping schedule of values for power either made or saved that is allowed to rise as high as $3,500 per MWh
3. Requiring all those providing DR to buy or create a safety margin of supplies in case their power reductions fail to materialize in the hour in which they are counted on by the system
4. A plan for system operators to implement selective blackouts "in a nondiscriminatory fashion" in case system operators can't maintain the necessary balance.[3]

The resolution of these issues is demanding considerable effort on the part of federal and state regulators and the stakeholders in wholesale power markets, who are almost all seasoned utilities, large private power traders, deregulated generation companies, and large user groups. As the Smart Grid penetrates the system more deeply, similar issues will affect the distribution systems and individual customers, all strictly in the province of state regulation. This will demand enormous efforts and new skills from state regulators—a theme I will return to more than once as the rest of the volume unfolds.

Markets also work for measuring another part of the avoided system capital costs, the costs avoided by not building the generators no longer needed due to locally saved or made power. Several organized power markets hold auctions for additional generation capacity, and the prices set in these auctions for one MW of new power supply are a good measure of the value of one MWh of peak usage. This approach works, but as with all other electric markets, the rules and regulations involved are complex, controversial, and require ongoing regulator involvement. "Set it and forget it" deregulation is a nonstarter.[4]

Deregulated wholesale markets all over the world employ dynamic pricing in the sense that they allow spot prices to vary every hour, leading to prices higher during peak periods due to the higher cost of the market supplying on-peak. These higher and more volatile prices, which are often the source of consumer complaints, are of course more lucrative for many small and renewable energy sources, and therefore encourage more DG development in these markets. This leads the proponents of markets to emphasize that they do a better job of encouraging DG and the DG industry in turn to generally support competitive markets.

However, old-school generation deregulation isn't much of an option for now in the rest of the United States, and isn't an option for valuing the avoided capital costs of the rest of the electric system (i.e., transmission or distribution savings). What is worse, because the transmission system is an interconnected web where all plants share the same lines, attributing transmission savings to reduced supply is a very challenging analytic exercise. Alternative transmission grids are extremely uncertain in their ability to get sited or built or their costs allocated; we will see more about this when we look at the transmission system in Chapter 7. The benefits of avoided transmission can also be very location-specific, and they can shift quite a lot over hours, seasons, and years. Regulators have to decide whether they should pay generators the avoided costs measured by the day, month, or year; the accompanying volatility and uncertainty make the recipient's revenue stream very risky.

Ever greater difficulties apply to measuring avoided distribution costs, where avoided costs are even more location dependent. Some customers may be attached to a part of the distribution grid that is so new and so large that there will be zero avoided distribution costs if customers at the end of the line increase their self-generation or cut their demand. Other circuits may be nearing a very expensive upgrade, so that DR or DG that prevents the utility from having to expand that particular circuit is worth several hundred dollars per kilowatt—almost as much as some small generators cost. This sensitivity makes rate setting for DG and DR much more complicated and necessitates sophisticated new planning methods such as locational *integrated resource planning* (IRP).[5]

You might be saying to yourself, "I thought that if I put a generator in my house it just runs my meter backward, in effect giving me credit on my power bill 1-for-1 between kilowatt-hours bought versus generated." If you live in one of the forty-two states where utilities allow this practice, *net metering*, you'd be right.[6] In this case, utilities and/or regulators decided to solve the valuation equation in a very simple way—namely, they assume that the value of distributed generation equals your current retail price. It isn't the worst approximation in the world, but it completely ignores all locational differences and scale effects. It may be a good incentive policy but it is not a particularly accurate price signal.

Locational variability notwithstanding, avoided capital costs have a feature that will prove critical when we discuss future scenarios for the growth of DG. This category of avoided costs is lower for generators located on the high-voltage grid, highest for DG in the middle of the distribution system, and lowest when the generator is located at the customer's end of the line. In light of the fact that avoided capacity costs are, by far the highest relative value stream from DG or DR, this means that it is most valuable to install DG in the middle of the distribution system, not at the far end of the grid, that is, at your home or factory.

Hard-to-Value Benefits

No one in history has done more to promote the cause of decentralizing the electric system than Amory Lovins. Since his discoveries of the thermodynamic inefficiencies in large power systems as a young physicist at Oxford, he has spent much of the last thirty years chronicling the vulnerabilities of the large-scale electric grid and the benefits of DG.

Lovins's most extensive treatise on distributed resources is his 2002 work *Small Is Profitable*, produced with colleagues at his Rocky Mountain Institute.[7] The volume is simultaneously a textbook on power systems economics, a handbook of small generator technology, and an extraordinarily detailed exposition on the value of distributed resources.

Parsing the benefits of DG in minute detail, Lovins et al. arrive at a list of 207 separate benefits of distributed sources relative to large-scale supplies. Many of these benefits are in the category of avoided capacity costs, which we've just seen are measurable, albeit with some difficulty. Due to their economic importance and measurability, they are always embedded in market or regulated prices. The rest of the 207 suggested benefits are not so lucky. From the practical standpoint, some of these are redundant and others are nearly impossible to measure ("fostering adaptive learning"), but the taxonomy is an extremely useful guide to the diffuse DG and DR benefits that aren't included in the prices paid to small generators or DR providers.

Table 5-1 condenses these hard-to-measure benefits into several simple categories. For each category (row), I provide an example or two, explain why utilities sometimes see offsetting disbenefits, identify the beneficiaries from this particular type of benefit, and discuss how the current framework of DG rules incentivizes the benefit, if at all.

The table illustrates that there are many technical and economic features of the Smart Grid, DG, and DR that provide diffuse benefits to all system customers. As an example, benefits like the reduced vulnerability of the grid to terrorist attacks are hard to put a value on, but they clearly benefit our nation as a whole.

It falls to regulators and policymakers to say whether and how these diffuse benefits will be incentivized or rewarded, and to balance them against occasional disbenefits shown in the table. Tax credits, consumer subsidies, low-interest loans, and many other approaches are sometimes used to provide economic support for the DG benefits that can't be included in the price paid

Table 5-1. Hard-to-Value Benefits of Demand Response (DR) and Distributed Generation (DG)

Type of Benefit	Examples	Possible Offsetting Disbenefits	Beneficiaries	How Value Is Conveyed to DR and DG Providers?
Planning and Investment	Less risk of new plant cost overruns	Less certainty over number, type, and location of future DR and DG providers	All utility customers in the region or market	State and federal policies that encourage DG and DR over central supplies
	Less risk of overbuilding due to long lead times			
Security and Reliability	Resilience against cascading blackouts and terrorist acts	Greater cyber-security threats to smart grid	All system customers	Policies as in Row 1 above
Reduced System Operating Costs	More sources of regulation, ramping, and reactive power		All system customers	In deregulated markets, some ability to sell these products separately
	Lower resource or fuel price volatility		All system customers	Policies as in Row 1 above
	Better management of distribution system voltage		All system customers	Policies as in Row 1 above
	Fewer large-unit stops, starts, and idling periods		All utility customers in the region or market	Good DR and DG pricing will pay directly for measurable locational benefits
Environment, Energy Efficiency, and Social/Community	Greater use of dispersed renewable sources; land use changes	Economies of scale in large renewable projects	Owners of dispersed renewables and utility customers	Policies as in Row 1 above

Source: Adapted by author from A. B. Lovins et al., *Small Is Profitable* (Rocky Mountain Institute, 2002).

for DG power. Add continuing evaluation of these policies to the already long list of ongoing work for the regulators of the Smart Grid.

Digging up the Old Wires

For all its benefits, the Smart Grid will come with large costs, the most significant of which will be the physical replacement or upgrading of the old electric distribution infrastructure. As we saw in the last chapter, the grid is designed, engineered, and operated to send power in one direction only. There are very few places in the low-voltage distribution wires where significant amounts of power are injected into the grid. But as downstream power generators become much more common the electrical design and controls over the distribution systems must be reengineered to allow power flows in two directions. This is technically feasible, but it is a big engineering job that will inevitably raise local issues.

To get a sense of the challenge, think of a city in which all of the streets go in only one direction. The city decides that nearly every street will be changed into a two-way thoroughfare. Exceptions are possible if the city decides that it is just too difficult or expensive to make the switch on one particular street. It is not hard to imagine that the city will need to build dozens of new traffic signals, along with many new side roads, on and off-ramps, and other traffic management features.

The distribution system will require the same sort of revamp. Wherever power is put into the grid, hardware analogous to the fuse box in your home must be added to keep the grid reliable and safe. Some parts of the local wires will be costly to expand or reengineer because the wires are inaccessible. One of the features of the Smart Grid is that it will tend to set power prices much higher for delivery to congested or overloaded corners of the local grid. Homeowners in these high-priced corners will love to generate power and earn a high price for it, while those who must buy everything at that same high price because they have no roof space to mount solar cells will complain bitterly.

When the distribution company gets around to upgrading the congested, high-priced corners of the system, the high prices will suddenly be eliminated.

The homeowner who just invested a hefty sum on solar panels with the expectation that local prices would be high will now be the one complaining, while the homeowner without roof space will be relieved. This phenomenon is close to what is already observed in the parts of the high-voltage grid where locational pricing is used in the wholesale power markets. It is one of several factors that have significantly slowed the expansion of the high-voltage grid, and it may well have the same effect at the local level.

Standards and Cyber-Security

The full vision of the Smart Grid includes utilities sending prices to customers, and directly to their devices, and the devices adjusting themselves accordingly. In order for utilities to use thousands of these devices to balance the system, the utility will have to be able to turn at least some of the devices on and off in emergencies. System controllers will need to know that the device is off, or on halfway, so communication between the utility and its customer and their devices must be via a two-way network.

All of this will work a lot more smoothly if the parts of the system all use the same technical standards, so they automatically connect together, physically, electrically, and in terms of their communication language. Greater standardization will give utilities and their regulators much more confidence that the products they are investing in will work properly; they will give the same increased confidence to individual customers, consultants, vendors, and investors, whose investment and adoption decisions will undoubtedly be influenced by standardization and overall ease of use.

The need for a two-way network between power system controllers and ultimately millions of homes, businesses, and devices raises a plethora of thorny technical and regulatory questions. Every electrical appliance and every telephone offered for sale in America uses the exact same plug and connects to its network immediately and without effort. Realizing this degree of "plug and play" interoperability means that, among other things:

- Price signals generated by the utility must be in an electronic format that every device recognizes.

- A common language and format must be developed so devices recognize commands from utilities and their own home control systems with commands like *turn off* and *turn on*.
- Every smart meter must connect to every utility's communication network so it can send and receive information.
- Home or business control systems—computers dedicated to programming devices and distributed generators and storage—must be able to talk to the smart meter.

These are all technically achievable outcomes, but they are also likely to be made difficult by several factors, some of which are unique to electric power.

One factor that is not unique is that good industry standards take time—sometimes lots of time. You may not realize it, but technical standards of this nature are not set by government agencies, they are set by nonprofit, industry-managed organizations such as the Institute for Electrical and Electronics Engineers (known as IEEE or I-triple-E) or the International Electrotechnical Congress.[8] These organizations convene committees of engineers who wrangle, often for years, over how to write a standard.

Of course, every piece of equipment and every technical process utilities use today is already the subject of numerous standards. The Smart Grid is such a sweeping development that impacts the entire utility operating chain, from power plant to customer meter. The following picture, limited only to communications networks along this chain, shows the numerous competing and overlapping standard protocols already in use today—twenty-five, plus or minus a few, shown in Figure 5-1. To try and speed up the Smart Grid, the Obama administration tapped the National Institute of Standards and Technology (NIST) to accelerate the standards-setting effort.[9] Following a U.S. Senate hearing on their progress, one technology blogger wrote of NIST's testimony: "There are so many standards to consider, said Patrick Gallagher, deputy director of NIST, that his organization's primary responsibility is simply prioritizing the order in which standards should be developed."[10]

While the standard-setting organizations jockey for the right to lead each standard-setting effort, firms that make devices are losing no time attempting to establish their technologies as the de facto winners. Often this is done through

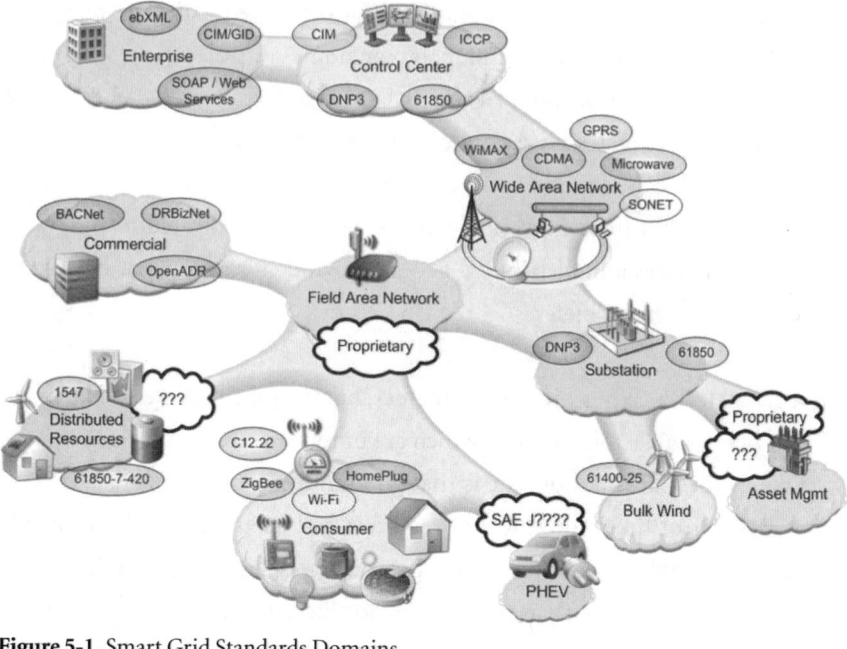

Figure 5-1. Smart Grid Standards Domains.
This figure shows each part of the electric power system and the current standards that already apply to subsystems and technologies that will eventually be integrated into a single interoperable Smart Grid. For example, the top center of the figure shows the utility-operated-control center. Four types of standards are common in control centers: CIM (Common Information Mode), ICCP (Inter-Control Center Communications Protocol), DNP3 (Distributed Network Protocol), and IEC 61850 (International Electrotechnical Commission), each with its own supporting organization(s) and history.

The control center will communicate with the equipment attached to the grid via a Wide-Area Network (WAN), shown below and to the right of the control center. Depending on the WAN technologies used (Wimax wireless, microwave, cellular CDMA, etc.), there are multiple standards to integrate. The WAN will communicate with the utility's field network, which uses proprietary technologies, standards, and substations, which use two of the same standards groups used by control centers. The number of domains the Smart Grid touches and the number of different standards already used in these domains indicate the size of the challenge of creating a single set of Smart Grid interoperability protocols.
Sources: EnerNex Corporation and California Energy Commission, used by permission.

alliances of manufacturers, software companies, and others, who set their own interoperability standard. For example, a consortium that includes Intel, Clearwire Corporation, General Electric, and Google have a system that uses a wireless wide-area network known as Wi-Max for data transfer; another group that claims over three hundred member firms advocates Wi-Fi technology as best for the Smart Grid.[11] Another consortium headed by Cisco, which makes equipment for the Internet backbone, wants to make sure that all data uses Internet protocol data packets.[12]

To cite another example, the push to accelerate Smart Grid standards is prompting the engineering community to revisit the technical requirements for downstream power sources—requirements that make it more difficult to use these sources as backup power. Right now every state has different technical rules for how generators can physically connect to utility grids, and many utilities have special rules themselves. Similarly, state regulators are looking at their DG policies, as when the state of New York recently raised the maximum allowable size for a customer-owned generator receiving the best terms almost tenfold, from 25 kilowatts to 2 megawatts.[13]

As Figure 5-1 shows, within any one utility, there is not a single control network with one language—there are a series of networks, each with its own specialized language, role, and interfaces to the other systems it must talk to. Utilities are extremely concerned about tampering with these systems. They are the backbone of its hour-to-hour balancing function, their response to transformer and outages large and small, and pretty much everything else. Yet these systems are not standardized. The systems utilities use to monitor and control their own distribution substations—the last point on their system they control before the wire to your house—are known as supervisory control and data acquisition systems (SCADA systems). Every utility has a SCADA system, but Sandia Labs expert Garth Corey notes that there are more than a dozen different languages and protocols used in the systems now in commercial use.[14]

To top it off, some officials in the U.S. Department of Defense (DOD) and some in the private sector are starting to express deep concern over the increased vulnerability of a Smart Grid to cyber attack. If devices can talk directly to power controllers, this might create a digital pathway for anyone with a home computer to hack in to power systems controllers. One company, IOActive, already claims to have invented a worm that can disable smart meters, and urges a slowdown in standardization and deployment efforts until security issues can be addressed.[15] In March 2009, the unofficial but self-proclaimed DOD Energy Blog reported that

Tom Donahue, the CIA's top cyber security analyst, made some news when he disclosed that cyber attackers have breached the electrical systems of multiple countries and have gone as far as powering down entire cities when their demands weren't met.

Of course, it's not business as usual for the CIA to speak out so publicly:

> "The CIA wouldn't have changed its policy on disclosure if it wasn't important. Donahue wouldn't have said it publicly if he didn't think the threat was very large and that companies needed to fix things right now."

But it's not just that hackers are getting more organized and more powerful. Grid breaches are occurring because new IT and communications technologies are making life easier for operators . . . and at the same time, more dangerous for customers.[16]

One month later the *Wall Street Journal* ran a front-page story claiming that both the Chinese and the Russians had already hacked into the U.S. power grid and were "mapping it" for possible damaging acts in the event of a confrontation.[17]

No one who can talk about it knows how much the need for security will seriously impede the Smart Grid's deployment. Whereas cyber security is certainly an issue in all computer networks, the level of concern and control is likely to be much higher for the Smart Grid than for many other applications.[18] It is nearly certain that Smart Grid standards will ultimately lead to a far more secure grid, but it is also quite possible that the need for security will significantly slow the deployment of key elements of the system.

Matching Benefits and Costs

Our ability to adopt effective and accurate regulatory processes for paying DR providers and distributed sources will be an absolutely critical factor driving the progress of the Smart Grid and the location of future supplies upstream or down. Good regulation will accurately assess the economic costs and benefits of distributed versus centralized supplies, grappling with the issues of timeframe, uncertainty, equity, and locational value. Historically, inquiries of this nature have taxed the resources of even the best regulatory commissions, a point I return to at the end of the book.

While regulation is steadily improving the methods of measuring avoided costs and building them into the payments to DR and DG providers, and examining other "barriers" to DR and DG, it is a slow process. Above all else, these regulatory challenges will determine how fast and how well we move into a future of interactive power systems unlocked by the Smart Grid.

CHAPTER SIX

The (Highly Uncertain) Future of Sales

WHEN I WAS A CHILD, my mother's kitchen had three electric appliances in it—a toaster, a mixer, and a fridge. Today, my kitchen has a microwave and convection oven, three laptops, five phone chargers, a CD player, and nine cooking appliances, plus or minus a few.

According to Michael Simonovich at the California Lighting Technology Center, "The average American kitchen uses 700 watts for lighting today," but his Center's test kitchen is designed to change that. It is a perfectly typical American model: gleaming white melamine cabinets, faux marble counters, and recessed lighting throughout. The kitchen is lit, brilliantly and evenly, with a crisp white light, a kind I associate with halogen lamps rather than the advanced LEDs (light-emitting diodes) that are actually installed. Mike tells me, gesturing at the room, "This kitchen uses 100."

Smart Grid or not, most of us tend to think that power demand must be destined to rise steadily in the coming decades. It especially feels this way as we connect to media and communications networks via an ever-larger number of "apps" and devices, all of which run on electricity. And for the emerging economies of the world this is true—electricity use is rising at a very rapid pace.

Yet despite the proliferation of devices and bandwidth, the United States and Europe have become mature power markets, with surprisingly modest forecasted growth. Power demand has become a race between new ways to use electricity and new ways to save power with all the devices we already have. There is no clear frontrunner in this race. The future sales trajectory for electric power has never been so uncertain, spanning a range from growth above current rates to gradual sales *declines* over the next several decades. This is especially true when you consider that the power customers make for themselves isn't part of industry sales.

The long-term trend in power sales has huge implications for the industry's future. If sales keep growing, power companies will have increasing revenues, which has generally made them financially healthy and able to raise capital easily. Regulated or not, it is easier to recover the costs of new supplies by adding them to power prices when sales are rising. Carbon emissions might be heading up with sales, but the industry has the fiscal capability to shift to low-carbon sources even if they involve high investment.

It is a different story if sales are declining. From the standpoint of the financial community, declining industries are not good candidates for new investment. The IOU part of the power industry has been viewed as a stable growth sector with high-dividend yields, allowing it to raise lots of capital at reasonable costs. In addition to changing the economic model and culture, a declining sales industry will have to raise prices frequently, as the costs of removing carbon emissions from existing sales will still be necessary. These costs will be spread over fewer and fewer kilowatt-hours, raising prices. Relations with customers and regulators will be unpleasant, with renewed agitation for deregulation as a means of escaping higher prices.

Driving Future Sales

Six major factors influence long-term sales, three of them positive and three negative. The positive factors are population growth, economic growth, and the trend toward electrification, especially in automobiles. The negative factors are higher electric prices, energy efficiency policies, and the onset of the Smart Grid.

Population growth is the most straightforward effect and perhaps the most surprising. According to the midrange U.S. census estimate, the United States will grow by over 100 million people by 2050. Over a million people immigrate into the United States a year; by 2050 it will be two million.[1] Even if we could keep per capita electricity use constant, as California has done for a decade, we would increase total power sales by nearly 33%. Distribution companies will also need to install about forty million more electric meters in new housing units and expand their systems accordingly.

Economic activity and power use are, of course, related. The stronger the U.S. economy grows, the more power is used by industrial and commercial firms and the more residential customers buy and use electrical equipment. Beyond the simple correlation of growth and power use, however, there is a secular trend toward electrification, that is, one not related to gross domestic product (GDP) growth. Overall, electricity is gradually stealing market share from other fuels for the overall mix of applications we use in the United States. In the residential sector, for example, electricity use is projected to grow six times faster than natural gas use through 2030.

During the next century this trend will take a giant leap forward. In the United States the largest use of energy outside the power sector is gasoline use for personal vehicles. As plug-in hybrid-electric vehicles (PHEVs) are introduced, electricity will gradually displace gasoline, boosting power sales at the expense of oil-based fuels. Other technologies will also migrate to electrification, though a few may migrate away from power toward natural gas or nongrid energy sources as well.

Over the long run, PHEVs represent a large new use of electricity. The timing depends quite a lot on how quickly these vehicles will become affordable and how well public policies encourage their adoption. To illustrate how widely PHEV market penetration estimates vary, the U.S. Energy Information Administration (EIA) forecasts that PHEVs will represent only about 2% of all vehicle sales by 2030[2] or only about 200,000 to 300,000 vehicles a year.[3] In contrast, a

2007 study by the Electric Power Research Institute (EPRI) and the Natural Resources Defense Council (NRDC) posits a midrange 2030 scenario with PHEVs achieving a 50% market share, or 7.5 million vehicles a year.[4]

The impact of these differences on electricity use is enormous. Electric sales generated by PHEV use according to the EIA forecast are insignificant; in the EPRI/NRDC scenario, we will need 282 million megawatt-hours—the output of thirty-eight large power plants—to "fuel" all these cars.[5] In the words of two industry consultants, "Rarely in history has an emerging technology offered such an attractive opportunity for the industry."[6]

Downshifting Sales

You might think that these positive factors would cause electricity sales forecasts to show robust growth over the next fifty years, but they don't. The numbers come with a bit of a story.

When President Carter created the U.S. Department of Energy (DOE), he wanted to make sure that all Americans—not just large energy companies—had access to accurate energy data. (Remember, this was long before the Internet made information so widely accessible.) As part of his energy program, he created the earlier-mentioned U.S. Energy Information Administration (EIA), the government's official energy forecaster. To keep EIA from appearing as if it was *advocating* new policies, its official forecasts were required to assume no future changes in energy policies beyond those fully enacted at the time of the prediction.

This prohibition on assuming policy shifts has great merit, but it tends to produce conservative results in times of rapid policy change. As of this writing, the Obama administration has the most ambitious energy policy agenda in modern times. As expected, EIA's official forecasts reflect none of the energy legislation in process.

Yet even with this built-in conservatism, EIA predicts very little electricity growth—about 0.89% a year. Residential consumption is projected to increase only about 0.8% a year; commercial energy growth clocks in at 1.7% a year. Remarkably, industrial electric usage is projected to decline in the 23 years from

2007 to 2030.[7] The utility industry's own forecasts, which are typically even more optimistic than EIA's, are only a little higher, or about 1.15% per year over all sectors.[8]

These rather tepid growth rates are explained largely by three forces that act to reduce sales. First, electricity is going to be more expensive over the next fifty years as the industry moves to low-carbon generation, expands transmission to reach renewable sources, and installs the Smart Grid. EIA's forecast, which does not include the impacts of climate legislation, say that prices will rise by 12% in inflation-adjusted terms by 2030. For example, the Environmental Protection Agency's (EPA) offered forecast of the impact of climate legislation on electric prices found an average annual rate increase of about 1.72%;[9] some estimates are even higher. As prices rise in real terms, sales will decline over the long term.[10]

Policies that accelerate energy efficiency beyond those embedded in EIA's forecast are another source of lower sales. Extensive experience with energy efficiency programs has shown that they are capable of having substantial impacts on sales. EPRI estimates that moderate increases in efficiency policies would cut the increase in total electricity sold *beyond EIA's estimates* by about 33% in 2030.[11] More optimistic estimates from organizations such as the American Council on Energy Efficiency Economy (ACEEE) suggest that stronger efficiency policies can reduce sales by more than 20%, reducing growth *below zero* over the next thirty years.[12]

EIA's official forecasts are barred from including the effects of stronger energy efficiency (EE) policies, including policies contained in proposed climate legislation. Considering the ambitiousness of the proposed policies, the history of measured efficiency program results, and a varied handful of forecasts of proposed EE policy impacts, it is easily possible that stronger policies could reduce EIA's already-modest sales growth by at least one-half. The real outcome will depend on the strength of the policy measures we adopt and our strategy for achieving them.

Finally, there are additional impacts from dynamic pricing and the greater control enabled by the Smart Grid. As we saw in the last chapter, dynamic pricing alone led to energy savings of a few percent, even without enabling technologies.

Although our ability to measure the savings potential of Smart Grid technologies is truly in its infancy, early studies show that energy savings from demand response and energy efficiency are in the range of 4 to 6%. At a national level, demand response technologies may reduce system peak reduction by as much as 4% by 2050. Studies have also shown that residential customers equipped with in-home displays can achieve on average 6.5% energy savings per device owner.[13]

Scenarios and Implications

Table 6-1 summarizes the six electric sales driving factors and the range of their impact on sales. The supporting calculations and sources are explained in the Appendix A for those who want to delve into the details.

As the first line of the table shows, U.S. electric sales were 3,725 billion kWh in 2008. EIA's no-policy-change forecast for 2030 is 4,527 billion kWh, an increase of 802 billion kWh over 2008. To this we first add 21 billion kWh for sales to more plug-in hybrids than EIA forecasts, since EIA does not account for the Obama administration's goal of adopting the policies necessary to put a million PHEVs on the road by 2015—more than triple EIA's current prediction.

To estimate the price impacts of likely carbon trading legislation, a renewable electricity standard, and other new policies, the table shows a somewhat

Table 6-1. Electricity Sales Possible: 2030 Scenarios (billion kilowatt-hours)

		2030	2008 Actual	Average Annual Growth Rate
	EIA Reference *Scenarios*	4,527	3,725	0.89%
Adjustments	Added sales from plug-in hybrids	+21		
	Reduced sales from higher prices	−86		
	Reduced sales from stronger energy-efficiency policies	−398		
	Reduced sales from Smart Grid– enabled technologies	−181		
	Reduced sales from expanded nonutility-distributed generation	−30		
	Net Adjusted Sales 2030	3,853	3,725	0.15%

conservative calculation of additional sales reductions beyond the EIA reference case. These estimated sales reductions, which come mainly from my own past work in electric forecasting, punch in at 86 billion kWh.

To account for EE policies stronger than those assumed by EIA, we use EPRI's forecast of "realistically achievable" additional savings beyond EIA's forecast, 398 billion kWh. This forecast is useful because it attempts to capture only the savings beyond EIA's forecasts, using traditional EE measures. It is also considered—by both its authors and its critics—as a conservative estimate of what could be achieved with strong EE policies. The EPRI study contained a more ambitious scenario showing 544 billion kWh savings with stronger policies. More recently ACEEE analyzed the very strong EE provisions of the Waxman–Markey climate bill and estimated that these provisions alone would save 578 billion kWh by 2030. For the sake of conservatism I use the more modest EPRI case of 398 billion kWh.[14]

The EPRI savings estimates included conventional *demand side management* (DSM) technologies but did not attempt to include Smart Grid–enabled EE. In a second EPRI study, three of my consulting colleagues, Ahmad Faruqui, Sanem Sercici, and Ryan Hledick, recently completed an extensive survey of Smart Grid–enabled EE savings. They found that Smart Grid technologies saved residential customers about 4 to 7% on their electric bill, apart from all other savings effects. To err on the conservative side, the table assumes 4% savings, or an additional 181 billion kWh.

Finally, oncoming electricity and climate policies are likely to trigger an increase in *distributed generation* (DG) not owned or marketed by utilities. As we learned in the last chapter, the growth of DG will depend strongly on regulatory policies that establish DG's total value (hence its price). To estimate the possible impacts of policies that promote DG, I rely on a special analysis that EIA conducted of the impacts of part of the first climate bill to pass the U.S. House, the Waxman–Markey bill.[15] This study found that the *renewable energy standard* (RES) provisions of Waxman–Markey would boost DG (i.e., reduce utility sales) by about fifteen billion kWh by 2030. I also consider an older EIA study that measured the impact of favorable tax credits and technology developments on DG. Although it is an old study that uses many outdated assumptions, it is indicative of the size of the policy-driven lift DG might receive over and above EIA's forecast. Based on all this, I assume DG policies could reduce sales by 30 billion kWh.

The bottom line of the table illustrates the billowing uncertainty surrounding future sales. EIA's reference case shows 802 billion kWh of sales increase through 2030, but the sum total of these adjustments reduces sales by 674 billion kWh—even when the administration reaches its PHEV sales goals. This leaves a net increase of 128 billon kWh or 3.4% in 22 years. This represents an average annual growth rate that is barely measurable—0.15%/year. While there may be some "double-counting" of sales reduction stimulated by higher prices, EE policies, and the Smart Grid, this illustration intentionally uses conservative estimates for each element in the table.

This exercise is not meant to predict the level of sales 21 years from now. Instead, it shows that, for the first time ever, one can create plausible scenarios in which sales go up handsomely and other equally plausible scenarios where they go down a bit, at least through 2030. Economic growth is a particularly important factor in electric sales. With economic growth at the same average pace experienced over the last 20 years (2.8%/year), instead of EIA's 2.4%/year prediction, 2030 power sales would increase by 12% percent over 2008 levels even with my adjustments. Conversely, low economic growth (1.8%) would leave electric sales about flat over the next thirty years, even with very small policy adjustments. In this case, the industry would face two decades of worse-than-stagnant sales.

The implications of this uncertainty for the industry are ominous. The power industry has very long planning cycles, and it is expensive to change or delay investments in new power supplies. Running out of generation has devastating consequences, and it is always necessary to have sufficient supply on hand to meet demand. If the trend in demand is variable or flat, however, planning decisions get especially complicated. No utility, and no utility regulator, wants to see a lot of money spent on a new power plant only to find, when it is finished five years later, that it is no longer needed.

As a utility consultant, I already see the impacts of this uncertainty in many of my clients. Utilities have a culture of building more plants and lines, and it is also the primary way that regulated IOUs make money. Yet I've never seen utilities as fearful of embarking on a construction program for a major plant as they are today.

Sales uncertainty is also important because it greatly exacerbates the rest of the challenges facing the industry. The scale of the investments required by the industry to convert its generation base to low-carbon sources, build out the transmission system and the Smart Grid, and contribute to energy efficiency efforts would be staggering even for an industry with healthy sales. Lower sales will make it harder (i.e., more expensive) to raise the capital to do all these things. Higher-priced capital and lower sales mean that the price of each of the lesser number of kilowatt-hours sold will be higher, further reducing sales.

A similar phenomenon was feared to occur in the power industry during the 1980s, when very expensive nuclear plants caused rates to rise quickly. The need to raise capital to finish nuclear plants with spiraling costs caused widespread financial distress, the largest municipal bond default in history, and the first utility bankruptcies since the Great Depression. The phenomenon became known as a "death spiral," where higher rates led to lower sales, which led to higher rates, and so on.[16]

My sales scenarios notwithstanding, I don't foresee anything approaching a death spiral for the industry, save perhaps one or two utilities facing odd and unfortunate circumstances. Having seen all this before, most of the industry's stakeholders understand that avoiding a financial train wreck is far better than trying to survive one. In other words, large new plants will not be built without fairly solid agreement between managers and customer representatives that they are needed. Still, sales uncertainty adds an unusually high level of risk to supply and capital planning at a time when the industry's investment needs are unprecedented.

PART TWO

Supply Side Challenges

CHAPTER SEVEN

The Aluminum Sky

FOR A BIRD'S-EYE VIEW of the attitudes Americans hold toward electric transmission, try perusing the energy blogs. One day about a year ago, SolveClimate.com featured a post on grid expansion that inspired many responses. One commenter named Mike wrote:

> If we want to implement a federal renewable portfolio standard, such as the 25% proposed by the Obama administration, or the current watered-down version of 15% [proposed] in the Senate, then we will need to look at both local and remote renewable resources.

Therefore, Mike concluded, we'll need to build a lot more transmission lines. Someone named Sheila also responded:

> To be clear, very *few* people in the "West" want our open spaces permanently destroyed and industrialized while our built environment bakes and sprawls and is not allowed to produce its own energy, just so Big Energy can, once again, profiteer on our backs. What *we* want (unlike our legislators who are owned by Big Energy) is

LOTS ... low interest loans for point of use solutions like efficiency and solar rooftops. ... The DOE determined, back in 2003, that 100% of the U.S. electricity needs could easily be met by using super-cheap thin film PV on existing rooftops. An additional 90% could be produced with the same material on in-city brown-fields. So, 190% of U.S. electricity needs can be met in the built environment with-out eminent domain, transmission-caused SF6 increases in global warming, water waste, dead ecosystems *or* wasted taxpayer and ratepayer dollars.

These two views are emblematic of a longstanding and often ferocious debate over U.S. transmission policy. At one end, new transmission lines are to be avoided at all cost, replaced in their entirety by energy efficiency and distributed generation. The alternative extreme is that the United States needs a vastly ex-panded national grid—some call it a national transmission superhighway—that enables nationwide power trading and better access to renewable resources. One environmental expert likened this to forcing America to live under "an alu-minum sky."[1]

Neither of these polar views is likely to occur, but they correctly recognize that transmission is the fulcrum on which the future of the industry will tip. The availability and cost of transmission will be a truly pivotal factor driving supply upstream or downstream, toward traditional (decarbonized) fuels or toward large, new, renewable generators. It is a fork in the road to the power industry's future.

Transmission Planning

The regulatory approvals needed to build a large new transmission line create one of the most difficult and time-consuming regulatory labyrinths in the en-tire utility universe. Federal officials regulate the rates and terms of service for high-voltage lines, including open access policies and all organizations that op-erate the grid. However, the federal government does not itself do any grid plan-ning, and it has almost no power to order lines to be built. Other federal agen-cies control access to federal lands and must give a number of environmental permits for most major lines.

The states retain nearly all of the power in transmission siting. A variety of siting and environmental permits are needed for a new line, most issued by the states. If a line passes through several states, each state follows its own permit process. And although the Federal Energy Regulatory Commission (FERC) has sole jurisdiction over transmission pricing, state public service commissions have a de facto veto over most lines because they control the ultimate recovery of costs from electric customers. They have many ways to penalize a utility that builds a line, or allocates the cost of a line, in ways they do not like.

Anyone asking for any of these approvals generally needs to be part of a transmission plan. Every region of the United States has a regional transmission planning organization. In areas where deregulation has taken whole or partial root—Texas, the Midwest, and the mid- to upper-Atlantic coast, there are nonprofit regional transmission organizations (RTOs) that lead regional planning, in addition to their role operating regional power spot markets and operating the grid itself. In the rest of the country planning is done by dedicated regional entities such as the Southeast Regional Transmission Planning group.

As we saw in Chapter 3, the transmission grid is like a network of water channels. Unlike natural gas pipelines or the Internet, power cannot be routed across its delivery network; it flows along the paths of least resistance from plant to load. Whenever a large new line is added, the flows in the region of the line rearrange themselves to find the new lowest-resistance pathways.[2]

This feature of the grid is essential, but it should not be exaggerated. First, significant flow rearrangement is important only for large new lines; small ones have minor impacts outside their immediate area. Second, the flow impacts of large lines diminish at a distance of one or two states away; only a small fraction of the total grid is affected. Third, lines that use direct current (DC), rather than the more common alternating current (AC) you have in your house, are different—they are controllable and are used for point-to-point controlled deliveries. Finally, there are some special technologies now in use that can control AC transmission line flows, and they will see greater use over time. All these factors make the impact of every large line somewhat specific to its attributes and electrical location in the grid (Box 7-1).[3]

BOX 7-1

Direct Current (DC) versus Alternating Current (AC) Transmission Lines

Electricity flowing in a transmission cable can flow in just one direction (direct current, or DC) or back and forth in tiny, rapid cycles (alternating current, or AC). Each type of current has its own pros and cons, and both have their place in a large transmission system.

When the goal is to build a "mesh grid" with many power plants interconnected through many lines to many load centers, constructing most of the system using AC lines is cheaper. On such AC grids, power flows can change direction freely, giving grid operators more flexibility to balance the system using many different combinations of power plants. It is also cheaper to change voltage levels. Higher voltage levels are more efficient over long distances, but wherever the power is offloaded to users the voltage must be reduced in several stages.

DC lines are more economical and/or functional than AC lines in certain applications. They are especially economical when a large volume of power needs to be delivered in one direction across several hundred miles or more. Changing DC to AC (to connect to the rest of the grid) is expensive, but you only have to do it at either end of the line.

DC lines are used to deliver hydroelectric power from the California–Oregon border to Los Angeles and from Canada into New England. These are also often better for moving power underwater and may be the best option for large offshore wind farms. DC lines are also the ideal links between two sections of an AC grid that need to be electrically isolated because they are controllable as to both direction and level of flow.

Nearly all transmission lines in North America are AC, but there are perhaps fifteen to twenty large DC facilities intermingled. As the demand for the transport of renewable energy from remote locations to cities goes up, the proportion of DC lines in the grid is likely to increase in the coming decades. New technologies such as superconducting cables, which can handle three or four times as much current as conventional wires, and new control technologies, will also heavily influence the future engineering and design of the grid.

In the planning entities, the planning process begins with proposals for new generating plants and new lines. Transmission planners, who are typically specialized electrical engineers, need to know—or rather, assume—where new plants will be built during the next ten or more years. Otherwise, their computer models can't simulate how power will flow in their region from plants to people. Proposals for lines come from utilities or sometimes from independent (but regulated) transmission line developers. There are also many opportunities to increase the capacity of existing corridors with better wires and other technologies.[4]

Utilities or independent owners of power plants who seek to build new plants must ask to be connected and are put in a queue. A wind farm developer might ask for a 600-megawatt connection to the farm he or she wants to place in service five years from now at a specific location. Because it is national policy to offer equal access to the transmission grid, any prospective plant builder with a little funding can apply for a place in the queue—there is no binding requirement that the plant ultimately be built. As a result, many builders apply for spots in the queue; only one out of every five or ten queue entries is actually completed.

Transmission planners look at their queue of "service requests" and estimate the power sources that will be in place five or more years out. They run many different scenarios and determine what parts of the regional network need to be expanded to prevent overloads and maintain reliability under a wide range of conditions. It is during these exercises that the impact of flow rearrangement from large new lines shows up in widely disparate places. They may find, for example, that a new line added in one state causes a transformer hundreds of miles away in another state to become overloaded. These exercises culminate in a proposed transmission expansion plan.

Remarkably, once they do all this work the resulting plan is merely advisory. No one is obligated to follow it, except in limited circumstances involving reliability. No one can be forced to build any of the new lines in the plan. In addition, planners don't have any siting authority, nor do they know which line will ultimately get siting approval, or on what timetable. Once again they make their best guess, but if the line they choose as their "preferred plan" subsequently runs into siting opposition and is delayed or canceled, it is quite literally back to the drawing board for them.[5]

You would think that this process would have created a strong coupling between transmission planners, utilities, and state public service commissions. However, deregulation and industry's multiple layers of jurisdiction create a structure that does not lend itself to coordinated expansion. In most states with retail choice, for example, utilities can't own or plan their generation, nor can state regulators—the deregulated market builds new plants. When power generators were first deregulated, no one thought to create a logical planning process for grid expansion, and we've been playing catch-up ever since. Even in areas where utilities are state regulated, the regional planning process is sometimes barely connected to state regulators and siting approvals.

In truth the reality isn't as bad as it sounds. Transmission-owning utilities are required to maintain reliability standards, facing penalties and embarrassment if they don't, so they are loathe to ignore any specific directive from their regional planners that clearly involves reliability. Because transmission is 100% rate regulated, the cost of this reliability-enhancing project is added into transmission rates as long as regulators agree. Regulators almost never refuse to allow rates to cover grid investments that maintain reliability, fearing the wrath of voters if blackouts occur. Quite a lot of what's in regional plans gets built, especially upgrades involving one or two systems at most.

The biggest expansion problems tend to arise when there are elements of the plan—especially brand new lines—that are not driven by reliability, but rather represent multiple options for providing new supplies with different economic and environmental consequences. For example, suppose planners find that there are two options for expanding the grid, one adding a line to a group of wind farms who want to sign contracts with local utilities and a second with a line into a forested area with biomass plants and a proposed new gas generator. Both options preserve reliability.

These options are sure to set off a heated discussion among nearly everyone involved in the plan. When this occurs, the planning engineers point out to everyone that they're just engineers. No one elected them to pick and choose the best option. Indeed, their transmission software only ensures that the grid is reliable—it doesn't even show which option is cheapest. To plan for lowest cost, you have to somehow integrate each utility's generation portfolio planning exercise—the exercise of picking among different generator types as discussed in Chapter 9. Transmission planning groups are just now starting to do this on a trial basis.[6]

Yet no one else has been chosen to make the decision either. Because the process is multistate, no one state regulatory commission has jurisdiction. State regulators control rates for distribution companies and generators if they are regulated. Public power and co-ops are subject to very limited FERC or state control. The FERC has full authority to set the transmission rates for whichever option is chosen, but it can only be reactive—it has no authority to pick the option in the first place, and it has zero authority to direct where generators are built.

With its limited authority, the FERC made a valiant attempt to improve regional transmission planning in 2007 by issuing Order 890. This order requires every private transmission-owning utility and RTO to adhere to a set of planning principles, including coordination with others, transparency, information exchange, and the inclusion of economic considerations. It was a big step, but broad-scale resentment against federal intrusion, legitimate regional differences, and its own limited authority meant that the new process is far from universal and actionable regional plans. More recently, the Obama administration has streamlined the process of federal transmission siting approvals, which will also help.[7]

Paying for New Lines

The multifaceted impact of a new line often leads to debates over who should pay for it, by way of its costs being included in some, but not other, utilities' transmission rates. There are several schools of thought on payment responsibility, or cost allocation as it is referred to in the industry. As its name suggests, the *beneficiary pays* school believes that the costs of new lines should be allocated to the ratepayers who benefit from the line. An alternative school believes that the value of new transmission is inherently shared by everyone in a region, as it is the network itself that provides reliable service to all, and that the environmental benefits of adding clean new supplies are enjoyed widely in the region. This is treating transmission as a regional public good, but more commonly this approach is (sometimes derisively) referred to as "socializing the costs" of new transmission or by the shorthand *postage-stamp pricing*.

One would think that it would be easy for economists to measure exactly who benefits from a new transmission line, but as an economist practicing in this area I can tell you that this a hugely challenging task. We can do it, but it takes a lot of assumptions about future prices (fuels and carbon), the success or failure of new plant proposals, and many other factors, all of which spark endless debate.

The results can differ vastly as well. Sometimes we find that the benefits are spread widely over many customers in a region, and sometimes they are limited to one part of the region. Often as not, the border between customers who gain measurably from the project and customers whose net benefits are uncertain doesn't match state or utility boundaries, making it hard to match the costs or benefits to particular groups of ratepayers. There are also legitimate philosophical differences on cost allocation, including questions about how to divide the value of greater reliability or allocate the value of better environmental outcomes.

The upshot of these difficulties and the many varied circumstances involving lines means that no approach works perfectly in every instance. This has stymied progress on creating a more routine approach to paying for lines, which in turn means that protracted debates over who bears the costs are the norm and not the exception for large new lines. The larger the line, and the more states it crosses, the longer and more ferocious the debate.

Because cost allocation is tied to the (cost-based) transmission rates that each company is allowed to charge, and transmission rates are federally jurisdictional, the FERC is the final arbiter of cost allocation disputes. However, because no one can force transmission owners to build a line, or the states to accept the FERC's allocation, the FERC usually must essentially seek a negotiated solution or watch the project collapse out of frustration and delay.[8]

Everyone in the industry is acutely aware of this problem, and some solutions are starting to emerge. Several RTOs have adopted default cost allocation policies. They usually fail to quell all controversy, but they are a start. In most regions and subregions, nearby states are starting to create organizations dedicated to regional grid planning and cost allocation agreements. In the Midwest, for example, the states involved in the Midwest Independent System Operator (MISO) regional market have several new fora, one for general discussions, one for debating cost allocations, and one for a smaller group of states that expects to build an especially large amount of transmission.[9] Similar groups and efforts are under way in other regions.[10]

The Need for New Lines

The convoluted process of adding large transmission lines wouldn't be too serious if we didn't need them. This brings us to the all-important question of how much transmission we really need.

In 2008 there was a little less than 40,000 MW of renewable capacity on the U.S. grid, not counting large hydroelectric dams.[11] There are already thirty-one states with *renewable portfolio standards* (RPS) mandates to install increasing amounts of renewable energy each year into the 2020s.[12] The aggregate effect of these mandates will be the addition of about 208,000 MW of renewable power by 2030—a fivefold increase. Perhaps by coincidence, studies show that this is also roughly the amount of renewable energy likely to be needed to meet the carbon emissions caps in the current legislation under discussion. Assuming these mandates aren't repealed, the only question is whether this capacity will come from small-scale sources (Sheila's dream) or large-scale plants that need transmission.

The potential sites for renewable plants needed to meet state standards and carbon caps are located widely around the country, but as we will see in the next chapter, their power is cheapest in the parts of the country where there are either strong winds, high isolation, or plentiful and cheap biomass, geothermal, and hydrokinetic resources. The National Renewable Energy Laboratory (NREL) has carefully mapped these resources and the results are no surprise: wind is strongest in the Great Plains and upper Rocky Mountain West, sunlight is strongest in the desert Southwest, and woody biomass is most plentiful in Maine and the rural Southeast.[13]

When you overlay a map of the large transmission lines in the United States over any of these renewables resource maps, you notice something obvious: the high-voltage grid wasn't designed or located to get power from these renewables-rich areas. It was created to move power from our current fleet of coal, gas, and nuclear plants, which are mainly located near fuel sources or bodies of water, to large cities and industries. Today, large transmission lines don't run out to the middle of the Dakotas or to the center of the Mohave Desert—why would they? There is also a high concentration of strong wind off many of the coastal areas of the country. Locating wind plants off coastal areas can produce very good wind energy, but obviously there aren't any transmission lines out into our coastal waters either.

In Chapter 5 we learned that balancing a grid with many large variable renewable sources requires additional balancing plants—demand response isn't yet sufficiently large or controllable, and large-scale storage is rarely feasible. These balancing resources need transmission too. Following a detailed look at the technical issues involved in relying on high levels of variable renewables, the group charged with maintaining U.S. grid reliability (North American Electric Reliability Corporation, or NERC) urged that new sources be spread out as widely as possible to take advantage of wind and solar diversity. It also said

> High levels of variable generation will require *significant transmission additions and reinforcements* to move wind, solar, and ocean power from their source points to demand centers and provide other needed reliability services, such as greater access to ramping and ancillary services [the correct technical name for what we've been calling balancing resources]. Policy makers and government entities are encouraged to work together to remove obstacles to transmission development, accelerate siting, and approve needed permits.[14]

The upshot of all this is that a low-carbon future requires much more transmission grid per kilowatt-hour than our current high-carbon grid. It is difficult to project the exact amount of new wire needed, but from a number of studies and conversations with planners I estimate that we will need an additional 30,000 to 40,000 miles of new lines through 2030. By comparison, the current high-voltage system is about 164,000 miles, and we've been adding about 1,000 miles a year during the last decade.[15]

Although simple math suggests that 1,000 miles a year times twenty years will yield 20,000 miles of lines, the reality is that this pace is much too slow. It takes five to ten years to site and build new lines when everything goes right, and generation must be planned in concert with the lines if they are going to be put to use immediately. Transmission line construction during the next several decades needs to be front-loaded to match industry policies, and that means a kick-start to transmission additions.

Fortunately, there is a clear upswell of transmission activities across the United States today. There are at least a dozen very large proposed transmission projects or proposals in the works, most of them designed to deliver renewable power to nearby states. Texas has approved a 2,300-mile series of lines to deliver wind from West Texas to the rest of the state. The Western Governors' Association is working on a massive planning effort, and several large new lines are well under way. Along the East Coast, a number of utilities and developers are proposing DC lines that will deliver onshore or offshore renewables from Maine and Canada into New England or offshore wind and other renewables northward from the southeast.[16] The latest figures for grid additions by all utilities in the next five-year period add up to 32,000 miles—more new lines than have been added in the last *twenty-five years*. It is doubtful that the industry can increase the pace of its line construction by a factor of five so quickly, but if something close to this unprecedented pace of line construction can be planned, financed, and built over the next several decades the grid should not be the constraint on our national policy objectives.

A National Transmission Superhighway?

Recognizing the need for transmission, economies of scale, and the concentration of cheap renewables in specific locations, some groups have raised the idea of building a national transmission supergrid often compared to the Interstate Highway System. One of the proposals, by the American Electric Power Company, would be a system of 765-kilovolt lines crisscrossing the country, as shown in Figure 7-1.[17]

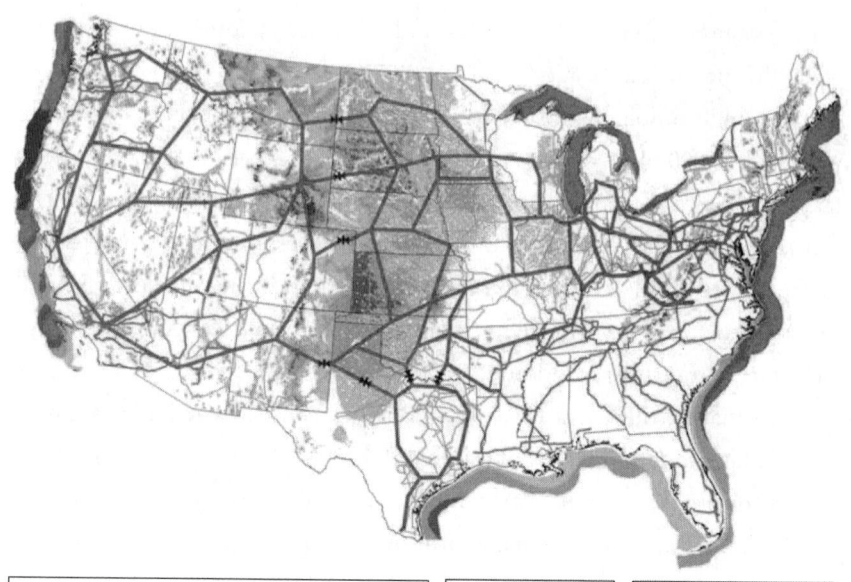

Wind Power Class	Resource Potential	Wind Power Density at 50 m W/m[1]	Wind Speed* at 50 m m/s	Wind Speed* at 50m mph	Transmission Lines Voltage (kV)	Conceptual 765 kV Network
3	Fair	300 - 400	6.4 - 7.0	14.3 - 15.7		
4	Good	400 - 500	7.0 - 7.5	15.7 - 16.8		
5	Excellent	500 - 600	7.5 - 8.0	16.8 - 17.9		
6	Outstanding	600 - 800	8.0 - 8.8	17.9 - 19.7		
7	Superb	800 - 1600	8.8 - 11.1	19.7 - 24.8		

Transmission Lines Voltage (kV)
- 234 - 499
- 500 - 699
- 700 - 799
- 1000 (DC)

Source: POWERmap
powermap.platts.com
©2007 Platts, a division of
the McGraw-Hill Companies

Conceptual 765 kV Network
- Existing 765 kV
- New 765 kV
- ►◄ AC-DC-AC Link

Source: American Electric Power (AEP)

*Wind speeds are based on a Weibull k value of 2.0

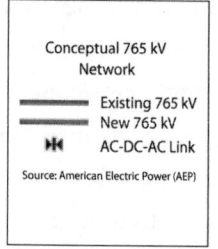

This conceptual transmission plan is designed to accommodate 400 GW of wind energy. This map shows the wind resource data used by the WinDS model for the 20% Wind Scenario. It is a combination of high resolution and low resolution datasets produced by NREL and other organizations. The data was screened to eliminate areas unlikely to be developed onshore due to land use or environmental issues. In many states, the wind resource on this map is visually enhanced to better show the distribution on ridge crests and other features.

Figure 7-1. Conceptual Plan for a U.S. Transmission Superhighway Overlaid on a Map of Wind Resources.

The shaded portions of this map show the areas where winds are strong and steady, which makes wind power economical. The darker the areas, the better the wind resource. The figure shows that most U.S. wind resources are located in the Great Plains region, from Oklahoma north into Canada, and off both ocean coasts and the Great Lakes.

Overlaid on the map is one concept of a national transmission superhighway system that would cover the entire country. Its function would be to transmit power from the Great Plains toward both coastlines, where most of the U.S. population lives. It would also send other forms of power long distances, including solar energy from the desert Southwest.

Source: This map was developed by the National Renewable Energy Laboratory for the U.S. Department of Energy, Department of Energy Efficiency and Renewable Energy. "20% Wind Energy by 2030: Increasing Wind Energy's Contribution to U.S. Electric Supply," DOE/GO-102008-2567, Washington, DC, July 2008.

Other groups have advanced similar ideas in the United States and also in Europe, where there is talk of a supergrid that will ship solar power made in the Sahara desert north into southern Europe—perhaps backed up by natural gas plants using North African gas supplies. Some of these plans rely on direct current (DC) lines, which as just noted are cheaper and more controllable when used to deliver power in one direction over very long distances.[18]

There are unquestionable scale economies in the use of high-voltage transmission, both AC and DC, to move large amounts of power a long way. One 765-kV line can carry as much as five 375-kV single circuits while occupying only about one-fifth as much land along its route (it does, however, use a much taller tower). Even with these scale economies, there is little evidence that we will need anything like a nationwide grid to achieve our policy goals.

The 30,000 to 40,000 miles of transmission we need is much smaller than a nationwide transmission highway system. Nearly all of the studies of interregional transmission options that measure total costs conclude that it is cheaper to use renewable energy within its own region of the country rather than ship it thousands of miles.[19] There is also quite a lot of opposition to a national supergrid from a variety of groups. Environmentalists' fears concerning transmission in general, captured in Sheila's blog entry, are only multiplied by massive national build-out.[20] Many local renewable developers and states also think that—despite the questionable economics of long-distance renewables—a supergrid will displace local resource development, depriving the region of new industries and jobs.

A national supergrid also raises extremely challenging reliability and security issues. While it is true that a more interconnected grid is inherently more reliable, this does not translate into a national supergrid improving customer-level reliability relative to the alternatives. Very long lines in a power grid can reduce reliability because they are harder to keep balanced and vulnerable to weather events and physical disturbances, including acts of terror. As the Smart Grid enters the picture, local distribution companies will have vastly improved reliability management tools and will rely less and less on the large grid for emergency backup. The entire question of grid design and the value of upstream versus downstream reliability enhancement deserve much more study, but it is already clear that we cannot equate a large, longer grid with greater reliability.[21]

Transmission Reform

Apart from those in the build-no-new-wires camp, there is general agreement on the importance of accelerating transmission additions in the high-voltage grid, provided energy efficiency and distributed generation are maximized first. Nevertheless, so far there has been little agreement on changes in federal authority designed to improve the present expansion process. Fearing that the federal government will site new lines across their boundaries, the states have been very reluctant to cede any authority over planning or siting approvals to the FERC or any other federal agency.[22]

In every piece of federal energy legislation, including all of the major climate change bills, transmission expansion reform has been one of the most controversial and hard-fought issues. Most governors and members of Congress don't like the idea of the FERC having the authority to overrule their siting processes, and they are also opposed to any legislative mandate on cost allocation because it might force their citizens to pay (albeit a very small amount) for grid additions that provide no perceived benefit. Although small increases in the federal government's authority to push grid expansion forward are quite likely, substantial new legislative authority is unlikely unless current plans start hitting road blocks. If current plans falter, all bets are off. The signs are that any federal mandates will have to be crafted surgically to apply under rather limited conditions.

It is much more likely that the regions of the country improve transmission planning, siting, and cost allocation due to pressure within their region from beleaguered utilities worried about meeting their mandates, renewable energy developers, and the environmental community.

The fear of greater federal authority may be another motivating factor. Whatever the cause, it is clear that the challenge is to finalize siting approvals and cost allocation for what is on the drawing boards.

The Tipping Point

Over the next decade, the rate at which transmission expands will do much to set the contours of the future industry. If a good portion of the current slate of large projects break ground, the momentum will continue and the present hearty rate of growth in large-scale renewables can proceed. This outcome will mean that the regions somehow found a way, whether through new federal policies or self-motivation, to cooperate enough on planning, siting and cost allocation to speed up the grid expansion process and make it less risky than in the past.

Realistically, less successful outcomes are entirely possible. Transmission expansion could falter for any of the customary reasons, not to mention some new or unforeseen problems. Another decade or two in which new transmission lags behind the need for new capacity will have dramatic consequences for the range of supply choices we can access. And now, more than ever, the more supply options we have the better.

CHAPTER EIGHT

The Great Power Shift

SINCE ITS INCEPTION, the generating sector of the U.S. power industry, and that of nearly every other nation on Earth, has relied heavily on fossil fuels. Thomas Edison burned coal in all his original power plants; by 1898, 50 tons a day were hauled through the busy streets of downtown Chicago to early power stations.[1] Today the United States still relies on coal for more than half its power; China 73% and India 75%. And coal use is rising worldwide. China alone is expected to commission 600,000 MW of new coal plants by 2030—more power capacity than is in all of Europe.[2,3]

As ever-more coal trains roll, the world scientific community is calling for an 80% reduction in greenhouse gas emissions by 2050, only forty years away. In developed economies like that in the United States, every study of this problem shows that electricity generators must be the first and the fastest to reduce their carbon emissions. During the next half-century, an industry that has rarely known anything but fossil-based generation will be forced to switch rapidly to low-carbon fuels. I call it the *Great Power Shift*.

While we know the shift is on, we don't quite know what the future supply picture will be. There are huge questions about the availability, cost, and reliability of many low-carbon-generating technologies. In the decade that will follow the present set of stopgap strategies, when the largest investments must be made, the uncertainties grow exponentially.

Le Tour de Plants

In the power business it is traditional to classify power plants by their size and their source of fuel. There are usually several different types of specific generating technologies that use each fuel, but in most cases the differences aren't as important within generators using one type of fuel as across generators using other sources. Size matters too, so let's begin with the larger sources and then move downstream to small-scale technologies and fuels.

Natural Gas

During the last twenty years, by far the most common power plants built in the United States use natural gas fuel. Gas combustion turbines are nearly identical to the jet engines on large airliners, except that airline jets use liquid aviation fuel. These engines are largely built in factories and shipped to a site, where five to twenty or more are formed into a single plant. By themselves, these plants are only economical if they run no more than a few hundred hours a year, so they are turned on only on days when power demand is very high. For this reason they are known as peaking plants, or *peakers.*

The second very common type of gas plant is a combined cycle generator. These plants have a conventional boiler that makes steam to turn a turbine-generator, just like the old science fair experiments. As its name suggests, however, a combined cycle gas turbine plant (CCGT, or often simply called a combined cycle, or CC) integrates two combustion turbines with the steam boiler in an extremely versatile and efficient arrangement. CCs have the highest energy conversion efficiency of any power plant—over 60% in some cases—and can run anywhere from 24/7 to a few hours a year with great controllability.

Natural gas CC technology has about the lowest cost and performance risk of any major power source. The technology has been proven, and it gets a little more efficient and cost-effective with each successive generation. The Energy Information Administration (EIA) predicts that the cost of building new CCs will drop from an already low $948 per kilowatt of capacity to $717/kW by 2030, with a 6% efficiency improvement as well.[4] Because natural gas emits less carbon than coal, and these plants are highly efficient, CCs have the lowest CO_2 emissions of any fossil generator, so their status as a low-cost resource is not easily threatened by carbon prices.

The main risk of relying on gas-fired generators is, of course, high gas prices. It is a very real risk. In recent years natural gas prices have careened between highs of $50 per million Btu and lows of $3.50, a range no other fuel can touch. At present, new supplies of gas extracted from shale formations in the United States are projected to continue to keep supplies ample and prices low for many years, but threats to our gas supplies remain.[5] For example, some gas drilling uses chemical fluids to fracture gas formations underground, and there is increasing concern that these fluids harm workers and leak into the water supply. There are also objections to increased access to public lands, among other issues.[6]

Utilities, regulators, and the grid's reliability managers also worry that the grid will rely too much on natural gas. Gas-fired power plants require an assured supply of gas, especially on the coldest days of the year, which is also when gas is most needed for home heating. Even though gas pipelines plan for such contingencies, a small mistake can have terrible consequences. During an especially cold spell in January 2004, barely enough natural gas reached New England to heat homes and fuel the region's power plants. Grid operators appealed for voluntary conservation efforts and narrowly averted rolling blackouts—a frightening prospect in the dead of winter.[7] Heavy reliance on gas also exposes power customers to price shocks when gas prices spike.

Low-Carbon Coal

Today coal-fired power plants everywhere in the world emit carbon dioxide in their exhaust gases. The technology to remove this CO_2 and store it away from the atmosphere is called carbon capture and sequestration (CCS). A variety of

CCS technologies are now being tested around the world, but there is not yet a guarantee that they will be ready soon for widespread commercial use.[8]

Looking forward, there will probably be three types of low-carbon coal plants. The first burns pulverized coal, nowadays often in "supercritical" or "ultrasupercritical" boilers with astonishingly high temperatures and pressures. A second, newer type of plant first converts coal to a synthetic type of natural gas, capturing many of coal's pollutants in the solid remnants, and then sends the gas into a modified gas CC plant. There are only a few of these new *integrated gasification combined cycle* (IGCC) plants in operation or under construction, but they are likely to become more common.[9] The third type of power plant, oxy-fuel combustion, adds oxygen to the boiler to make postcombustion removal of CO_2 easier.

IGCC *without* CCS is already in use at commercial scale around the world; the first commercial U.S. facility, a 630-MW Duke Energy plant in Edwardsport, Indiana, is scheduled to generate by 2012.[10] The main purpose of the U.S. Department of Energy's (DOE's) FutureGen project in Mattoon, Illinois, is to demonstrate how CCS can be added into IGCC plants to make a carbon-free IGCC, a technical challenge even its proponents acknowledge is substantial. Remarkably, China's own 650-MW IGCC demonstration plant *with CCS* is also scheduled to be ready by 2012.[11]

What about the 1,500 existing coal-fired units in the United States—and the thousands more around the world that exist or are scheduled to be built without CCS? American Electric Power and the Southern Company are each testing 20-MW pilot units from different vendors that remove CO_2 from a power plant's smokestack. In mid-2009, the DOE also announced funding for a 120-MW postcombustion demonstration at Basin Electric's Antelope Valley coal plant.[12] As with IGCC technology, the question is not whether the technology works, it is whether the cost and reliability of the process at commercial scale yield a competitive total cost of power as low as that of the competing options.[13]

Sequestering the CO_2 after it is collected is an entirely separate process. Carbon sequestration takes the CO_2 extracted from coal plants, in the form of a slurry or cold liquid, and injects it into depleted oil wells, underground saltwater aquifers, or other geological structures. While it may be surprising, most of the main methods have been in small-scale use or larger trials for some time now. So far, scientists think that there's a 99% chance that properly sequestered carbon will stay put for a thousand years.[14] Moreover, there appear to be more than enough sites with sequestration capacity to accommodate coal emissions for the entire next century or more.[15]

The most important questions about sequestration are regulatory. At present, no one is sure what legal liability the owner/operator of a sequestration site has if the site fails by leaking CO_2, perhaps catastrophically. Sequestration sites may also pollute groundwater or nearby mineral deposits or have other unintended harmful impacts. The legal framework established must assume that these sites will be monitored and maintained for hundreds of years.

However, neither the Environmental Protection Agency (EPA) nor any other federal agency yet has the authority to comprehensively classify and regulate sequestration.[16] Environmental law experts such as Michael Gerrard, head of Columbia University's Center for Climate Change Law, predict that the regulatory scheme finally adopted will then face many years of litigation in the courts before it becomes settled law.[17]

The system for sequestering CO_2 will also require a massive physical infrastructure that may not be feasible to finance and build, or may take longer than we can afford. The Massachusetts Institute of Technology's (MIT's) *Future of Coal* study points out that 60% of the current emissions from U.S. coal plants, if captured and converted to liquid CO_2, would represent a flow rate of twenty million barrels a day—about equal to our current national use of oil.[18] After studying the issue, three experts from the nonprofit Clean Air Task Force concluded that the pipeline network needed to sequester all coal-derived CO_2 would number at least 30,000 miles—but that this seemed achievable in view of the fact that the United States added 150,000 miles of natural gas pipelines between 1960 and 1980.[19]

Finally, we cannot ignore concerns over the environmental costs and availability of coal. It is often assumed that coal is abundant in the United States—we are sometimes referred to as the Saudi Arabia of coal—and that the environmental consequences of mining are controllable and publicly acceptable. While these considerations are unlikely to slow the growth of coal outside the United States, coal production in the United States may be reaching its peak, and both environmental concerns and rising costs create uncertainties for a fuel whose availability at low delivered costs was previously never questioned.[20]

A Nuclear Renaissance?

Depending on whom you talk to, nuclear power is either poised for a major renaissance or destined for another set of costly failures. The 100 U.S. plants currently operating were built between 1967 and 1990. They now supply 20% of the nation's power with high reliability and low operating costs.[21] These plants will all retire during the next fifty years, leaving a huge hole in our electricity supply from plants that do not produce greenhouse gases.

So far, seventeen U.S. nuclear power plants have applied for licenses— though U.S. nuclear regulators don't expect more than six to start construction right away.[22] Internationally, there were forty-one plants under construction as of June 2008, primarily in Asia, with China reportedly planning to build 100 by itself.[23] Touting the greenhouse gas and environmental benefits of nuclear power, the Republican Conference of the U.S. Congress recently released a plan to build 100 new U.S. plants by 2030.[24]

By a wide margin, the fate of nuclear power rests with its capital costs. The costs of everything other than plant construction—fuel, operations, waste disposal, and decommissioning—add up to only a few cents per kilowatt-hour.[25] The credit rating agency Moody's recently concluded that the "risks of building new nuclear generation are hard to ignore, entailing significantly higher business and operating risk profiles, with construction risks, huge capital costs, and continual shifts in national energy policy."[26] If new nuclear plants can be built for about $4,000 per kilowatt of capacity—roughly twice the cost of a coal plant and four times a gas-fired unit—they still produce competitive power because all their fuel costs are so low, even counting waste disposal. However, if construction costs $7,000 to $8,000/kW or more nuclear power is unlikely to be economical.[27]

At this point the new U.S. plants are on the drawing boards and cost estimates span precisely this range. EIA estimates that new plants cost under $4,000/kW, and most of the current companies building them predict costs in this area. EIA also predicts prices will drop significantly (to under $3,000/kW) as the industry gains construction experience, which is normal in the power industry. Many other analysts are not so sanguine. A 2008 Congressional Budget Office comparison of generating costs concluded that advanced nuclear plants would cost 30 to 35% more than coal- or gas-fired power.[28] Other experts, such as Joe Romm at Climate Progress, project nuclear costs far above the alternatives.[29]

Of all the large technologies, nuclear power also faces the largest supply chain challenges. Many of the parts in a nuclear plant, such as pressure relief valves or containment structures, must be certified to meet higher standards than other similar parts used in other generators. The materials used to make these parts are often of a higher quality, and therefore more expensive, especially in a period of rising commodity costs. Many of these parts are made by only a handful of manufacturers in the world with the right certifications.[30]

The state of the nuclear workforce is a closely related concern. Citing statistics from the Nuclear Energy Institute, two experts note that fully half the nuclear workforce will be eligible to retire within five years.[31] A recruiter for a nuclear firm explained to me that his company is simply unable to recruit enough new engineers to take the place of highly specialized retiring workers, in part because the number of U.S. nuclear engineering university programs has dropped (from sixty-five to fewer than thirty) and because there are restrictions on the foreign nationals they can hire. The Health Physics Society claims there is already a 30% shortage in radiation protection workers.[32]

Waste disposal also continues to be an issue. President Obama recently cut the funding for the only long-term nuclear waste repository in the United States, and there is scant progress at finding an alternative site. Waste is now being stored in temporary repositories adjacent to current reactor sites, raising the threat of release by accident or acts of terror. In the same vein, a single major accident is likely to greatly slow if not stop any nuclear revival in its tracks.

Onshore and Offshore Wind

Onshore wind power installations are becoming a common sight in rural America and every other windy part of the world. Modern wind turbines, ubiquitous in pictures of the new energy economy, produce up to 3 MW each and have rotors as wide as a football field. Wind farms, where turbines are mounted on huge steel poles every 30 to 60 acres,[33] can generate as much power on windy days as a large nuclear plant—though the average farm is more like a tenth this size (83 MW).[34]

The economics of wind power are well established, and the technology is entirely proven, making wind one of the least risky generation technologies around. Any large site where the wind blows at least 25% of the time at speeds above 9 miles per hour, and where there is nearby transmission able to absorb the wind power at that spot, and where siting permission can be obtained is a good place for a wind farm. Under these conditions, wind power costs about the same as power from natural gas plants, not counting the costs of transmission and grid integration. As technology improves, the cost of wind is expected to drop at least another 10 or 20%.[35]

No surprise, then, that onshore wind has become by far the largest and fastest growing renewable energy source in the world. In 2008 the United States added over 8,000 MW of wind capacity; China was in second place with another 7,000 MW. The National Renewable Energy Laboratory long ago mapped the entire country's winds to determine the number of sites with economical wind conditions. Their estimates of an economical wind resource immediately led wind proponents to dub the United States the Saudi Arabia of wind and has recently been estimated at 8,000 GW of available land-based wind resources.[36]

While no one questions the potential, wind has its own unique challenges that temper its growth. The largest factors limiting wind development are unquestionably the availability and cost of transmission. Wind power is much more economical where winds are strong and steady, and (as shown in Figure 7-1) the most economical wind locations in the country are deep in the Great Plains and off the U.S. coasts.

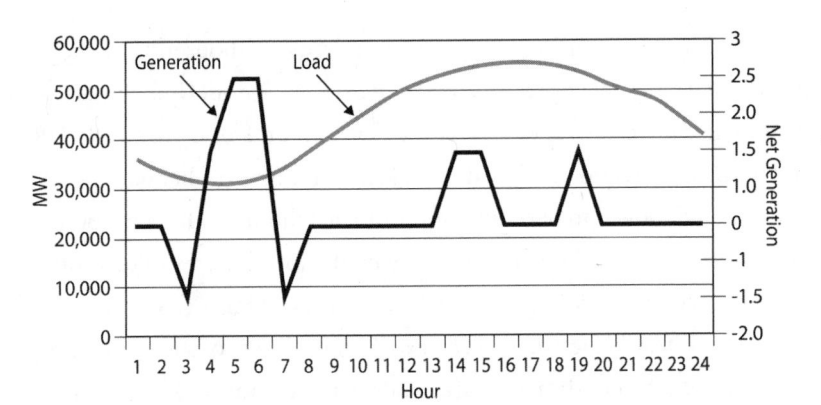

Figure 8-1. Comparison of Time Profiles of Total versus One Wind Farm's Output, July 18, 2008.

The smooth upper line labeled load is the approximate total hourly demand for the entire PJM-East (mid-Atlantic) power market on this date. Demand follows a typical pattern of falling to its lowest level in the hours between midnight and dawn and rising to its highest levels between noon and 8 p.m. (Hour 20).

The mostly lower jagged line shows the net flows from one wind farm selling to the Exelon Generation Company, LLC, in PJM-East. Net flows equal the electrical production of the wind farm minus energy consumption at the windfarm. These net flows are highest between the hours of 4 a.m. and 6 a.m., when total PJM-East demand is at its lowest point. There are two helpful bumps in wind generation between 1 p.m. and 3 p.m. this day and around 6 p.m. However, overall there is a large mismatch between the timing of wind generation and systemwide demand on this particular day.

Source: (c) 2009 by Exelon Corporation. All rights reserved. Used by permission. This work may not be copied, used or distributed in any way without prior written permission of Exelon Corporation.

Wind is a variable resource that produces power on its own schedule and can't be controlled by system operators to match supply and demand. Figure 8-1 contrasts the actual output of one wind farm in Pennsylvania to overall power demand in that region on July 18, 2008. As the figure shows, it is not uncommon for wind to blow most steadily at night, when power is needed least.

Earlier we saw that one of the Smart Grid's best features was its inherent ability to balance variable sources by shifting demand around (demand response, or DR), harnessing diversity, and installing electricity storage. This will all come to pass, but the Smart Grid is largely conceptual now, whereas utilities already rely on thousands of megawatts of wind.

Balancing the system on a scale of hundreds or thousands of megawatts must be done by system operators on the high-voltage grid. Unfortunately, these operators usually can't yet count on massive amounts of controllable DR that could balance generation and load. That leaves storage. In Chapter 5 we saw that storing enough power to partially supply one neighborhood might soon be feasible and economical. Unfortunately, storing the multimegawatt quantities grid operators need to buffer large wind farms is a much bigger challenge and is not expected to be feasible for many decades. (Storage is one of those rare technologies where costs do not drop at a large scale—at least not yet.)[37]

With no other current options, grid operators continue to balance the system by turning controllable plants on and off. For this reason, the cost of power from one wind farm alone is not a fair measure of the costs of electricity supply from wind; you must also count the costs of complementary power from backup sources as well. When these "grid integration" costs are counted, the blended cost of wind and backup resources is roughly 30 to 50% higher than the cost of wind power alone.[38]

Due to the technical demands of keeping grid balance, current engineering studies suggest that it will become prohibitively expensive and/or unreliable to use wind for much more than 30% of the system's total power needs, at least for now.[39] These limits are a complex function of the amount of transmission available and many other attributes of the region's overall power system. This issue is undergoing intensive study and debate, and the industry may well find ways to achieve higher levels of wind reliance, especially if it can build more transmission lines.

Offshore wind plants are common in parts of Europe, and the United States has just begun to lease offshore sites for wind development along its own ocean coasts. Offshore winds are strong and steady, but turbines that sit as much as 18 miles offshore are much more expensive to install and maintain, transmission is expensive, and environmental permitting can be difficult. Offshore wind is unlikely ever to be anywhere near as cheap as it is onshore—construction costs are roughly double—but it may well play an important role in regions where state and local officials want to encourage carbon reductions and energy self-reliance, even if it costs a little more.[40]

Photovoltaic Solar

Photovoltaic (PV) panels, also known as solar cells, convert sunlight directly to power, and are seen on many large roofs today, especially in Spain, Germany, and California. The two basic classes of commercial PV technologies are silicon-based flat panels and so-called thin-film technologies. Each has its own advantages and market applications.

Silicon panels are the rigid, blackish-blue, glass-covered panels seen on many rooftops, often in standard 3-foot by 4-foot sections.[41] These panels have steadily improved their efficiency and lowered their cost. When PV power began in the 1980s, it cost over $1 per kilowatt-hour, too expensive for anything but spacecraft.[42] Today it is down to $0.19 to $0.32/kWh—still well above most other sources, but only one-fourth the original cost.[43] Meanwhile, the average conversion efficiency for these cells has increased steadily, from 3% initially to over 18% today.[44]

The second type of cells, now with about 10% of the market, are *thin-film* technologies.[45] These cells are made by depositing semiconductor materials on flexible substrates, such as sheets of plastic roofing material. The most common semiconducting material is a combination of cadmium and tellurium, but other chemical combinations are also used. Thin-film cells are cheaper to make and more versatile, as they can be integrated into roofing materials, siding panels, and even windows, where they are nearly invisible but still make power. However, thin-film cells are less efficient, so the power produced can be more expensive per dollar invested.

Among all power-generating technologies, PV cells are the closest to a manufactured, mass-produced product. PV panels are made in enormous factories quite similar to semiconductor plants, with very high volumes. As with much high-tech manufacturing, new facilities with lower costs are opening across Asia; China now reportedly has more than 100 PV manufacturers.[46] As cell production increases and disperses, continued competition and cost efficiencies are likely to reduce cell costs.

Unfortunately, however, PV installations are more than cells. PV panels of all types need land, mounting systems, inverters that change the cells' direct current output to alternating current, and a fair amount of wiring. These costs, known as balance-of-system (BOS) costs, now almost equal the costs of the panel—about $1,500 per kilowatt.[47] "I can buy an entire gas-fired power plant for what I pay for the balance of system," grouses Bob Hemphill, president of the worldwide developer AES Solar. Unlike mass-manufactured panels, BOS costs are unlikely to decline so quickly. Still, total PV costs are expected to decline to $3.50/watt, or about 13 cents per kWh, in the next decade or two—well within the range of other power sources.[48]

As was true for wind farms, PV power varies with the weather and must be paired with long-term and short-term backup supplies or storage to maintain grid stability. This adds roughly the same magnitude of costs as wind requires for grid integration and control, pushing the true average costs of PV power into the range of 30 to 50 cents per kWh. Despite these high costs, its versatility and environmental quality have made PV quite popular. There are a number of federal and state tax credit and other subsidy programs, some of them reducing the net costs of systems dramatically. With these aids, 342 MW of new PV cells were added in 2008 in installations as small as a single rooftop and as large as a 13-MW plant at Nellis Air Force base in sunny Las Vegas.[49]

Concentrating Solar Power

As the name suggests, concentrating solar power (CSP) plants concentrate sunlight with reflectors and use it to make power. These plants, sometimes also called solar thermal electric, appeared briefly in the 1980s; an improved cohort is now under construction in the United States. Unlike PV, this is strictly a large-scale operation; one plant generates 80 to 200 MW using thousands of acres of land.

There are two types of CSP plants entering commercial operation today. The first uses long rows of parabolic mirrors to boil a special kind of fluid and uses the resulting steam to turn a turbine generator. There are a handful of operating parabolic plants in the United States totaling 419 MW, several under contract, and one under construction in Florida. The second type uses mirrors that focus the sun on a single "power tower" where a working fluid (currently water) is boiled at very high temperatures.[50] There are not yet any U.S. operating power towers (an experimental unit operated briefly in the 1990s), but one company, BrightSource Energy, already has over 2,600 MW of new plants under contract in the United States today.[51]

CSP plants require very high levels of sun and quite a bit of land, so they are now economical, mainly in the desert Southwest. However, within that area there is a very large solar resource, and many developers have scrambled to get siting approvals for new plants they would like to build. Because much of the land in this area is owned by the federal government, the siting work will proceed through the U.S. Department of the Interior's Bureau of Land Management (BLM). Remarkably, BLM reports that it has *already* received 158 applications for 97,000 MW of new CSP capacity—roughly one-eighth of all capacity in the United States—covering 1.8 million acres.[52]

These astonishing numbers notwithstanding, CSP is still not a fully proven technology and it faces its own unique impediments. CSP plants produce only during the day while the sun shines brightly, which is a valuable time to make power but still requires backup and grid integration like wind and PV. In contrast to wind and PV, however, the fact that CSP uses a working fluid makes it possible that these plants can store solar energy as heat and then generate power for a while after the sun goes down. This sort of internal storage will make CSP plants close to equivalent to dispatchable peaking plants and will lower the grid's less visible integration and backup costs.

Transmission is another large uncertainty. As we saw in Chapter 7, the western U.S. grid was not designed to move large amounts of power produced in undeveloped parts of the desert. While the federal government and western states are setting their sights on changing this, it will take many years before the results are known. Finally, some of the current CSP designs use large amounts of water, which is obviously scarce in any desert area. So-called dry cooling, in which no water is used for cooling, reduces water use extensively, but it also makes plants less efficient and more expensive to build.

Assuming the technology works as expected, these plants should produce power for 10 to 15 cents per kWh before the inclusion of backup or storage costs. Estimates for the cost of added storage are speculative at this point; one analyst estimates about 1 cent per kWh, an increase of about 10%.[53] For plants that need it, dry cooling is estimated to cost another 10 to 20% as well.[54] Since CSP is relatively new, these costs should come down over time. EIA, for example, forecasts costs dropping nearly 50% between 2010 and 2030, about as fast as PV declines.[55] For a technology that is already nearly competitive and has storage and controllability, this represents a pretty attractive package.

Biomass Power

Biomass power is generated from four different streams of biomass fuel materials: wood wastes from paper and furniture manufacturers, forestry residue, agricultural residue, and methane from municipal solid waste ("landfill methane"). The most advanced biomass plants use the same integrated gasification combined cycle technology (IGCC) technology as modern coal plants to improve efficiency and reduce air pollution. Biomass power plants in the paper and furniture industry are also often cogenerators, as these industries have high needs for process heat.

The actual generation of biopower emits just as much CO_2 as any other similar fossil-fueled plant, except for landfill methane. When dedicated crops are grown for the facility, the atmospheric carbon absorbed during crop growth roughly equals the amount emitted during generation. This particular arrangement, known as closed-loop biomass, is considered net-zero carbon.[56]

The capital costs for biomass plants are similar to those of coal plants, so biomass has tended to be used where plentiful, high-quality supplies of fuel cost the same as or less than coal and where the waste heat can be used.[57]

Geothermal Energy

Geothermal power plants use ultrahot water from deep inside the earth to make steam and generate electricity. Where natural formations allow easy access to naturally superheated water, geothermal power technologies are a proven, reliable, cost-effective source of baseload power.[58] The constraints on this technology are entirely related to the availability of viable natural sites with nearby transmission capacity.[59] Currently there are about 3,000 MW of geothermal power located entirely within four western states, Alaska, and Hawaii. Technical advances could enable much broader use of geothermal heat in future decades; the current resource and technology base is projected to provide a small but significant zero-carbon contribution of about 25,000 MW by 2025.[60]

Hydroelectric and Hydrokinetic Energy

The industry has pretty much given up on building large new hydroelectric plants in the United States (though not in Canada and some other parts of the world). However, there is a new generation of "hydrokinetic" technologies that make power from river and ocean currents and ocean waves. According to the blog cleantechnica.com, at least one small riverine turbine is now submerged in the Mississippi River near Minneapolis.[61] The British and Danish governments are also both sponsoring small new test machines that produce power from wave energy, and the British recently announced plans to build by far the largest wave energy machine in the world, a 20-MW unit off the coast of Cornwall.[62] These systems are still untested and far from commercial, but they hold the possibility to supply vast amounts of low-carbon power in the distant future.[63]

Distributed (Small-Scale) Technologies

One bit of confusion sometimes muddles discussions of distributed sources. Sometimes analysts treat *all* wind and PV power sources as if they are small, decentralized sources. This doesn't work for capacity planners, who need to distinguish between large- and small-scale renewables, or equivalently, centralized and distributed generation (DG). In this chapter, I follow the California Energy Commission's definition of DG as power sources less than 20 MW connected to the local distribution wires rather than the high-voltage transmission grid.[64]

There are four main electric technologies in DG's future: combined heat-and-power (CHP, or cogeneration) plants (also called microturbines or small cogeneration), small wind installations, small PV plants, and fuel cells.[65] Solar hot water heating, daylighting, geothermal heat pumps, and other similar technologies use renewable *energy* to cost-effectively *displace* the need for electricity. Because they avoid the wasteful process of generating power and instead use renewable energy directly, I don't treat them as sources of electric power.[66]

The most common clean distributed source is CHP, which uses the waste heat produced by electric generators to heat buildings and industrial processes, squeezing far more useful energy out of a Btu of fuel than most non-CHP power plants. As a result, end users' total energy costs are much lower than they would pay to buy heat and power separately. In order to use the waste steam from a plant you need a heat-user (known as a steam host) that is the correct size to absorb the surplus steam, located close by, since steam can't be piped very far. Plants in the 5 to 20 MW range are well matched to the steam needs of factories and large commercial or residential complexes, making this most commonly a medium-scale technology. In New York City and some European and Russian cities, there are networks of steam tunnels to many buildings that allow for the use of waste heat from utility-size plants. Unfortunately, because creating these tunnel networks is prohibitively costly, the United States is unlikely to expand cogeneration on this very large scale.

The reasons why industrial-scale CHP has not been used more widely in the United States are all related to regulatory and institutional hurdles. It can be difficult to make the necessary coordination arrangements with a large building that will accept and use a generator's waste heat. It also requires navigating many siting, land use, and other rules to put generators into or near heat users. Arrangements with utilities are also a frequent issue. Because cogenerators displace utility sales, utilities don't have an economic incentive to help them get established—yet utilities have to connect up and monitor the cogenerator and provide backup service when the cogenerator trips off (some cogenerators are "off the grid," in which case there is no backup, but most are not).[67]

Another new development spurring the growth of this technology is the commercialization of home-sized CHP units, or microturbines. These tiny, gas-fired power plants sit outside a home or business, make power, and send the waste heat in to supplant the furnace or other heat sources. After many years of development, several vendors now sell these units to homeowners and property developers.

Power from these units is still much more expensive than large-scale cogenerators, though it is cheap compared to other small sources. Hooking them up also involves locating heating, cooling, and electrical equipment much more carefully than in the average building. On the other hand, because microturbines are new to the market, EIA forecasts that they will become much cheaper over the next two decades, to the point where they will be an extremely economical option for most new buildings if—and this is a big if—natural gas prices stay moderate.

Small-scale wind installations are the third most common distributed power source.[68] Residential-scale turbines are units up to 100 kW in size, designed to work with very little control or maintenance at medium wind levels (9 mph average). Their power isn't nearly as cheap as large-scale wind—12 to 15 cents/kWh versus 4 cents, excluding backup and integration—but it is one of most economical power sources where the grid isn't available ("off-grid").

Despite its high on-grid cost, there are twenty-two manufacturers of small turbines in the United States alone,[69] and they are optimistic about their market prospects, having experienced an eye-popping 53.8% growth between 2007 and 2008.[70] This will be aided greatly by technology-driven cost reductions; EIA expects small turbine prices to drop 33% in real terms by 2030.

The final distributed technology is fuel cells. Contrary to an impression often conveyed in the media, fuel cells are not pollution-free electric generators; they are a natural gas–fueled unit that makes power using an internal chemical process, like a battery in reverse. The future may hold fuel cells that use pollution-free renewable sources, but for the foreseeable future natural gas fuel cells will emit about the same pollutants as gas-fired microturbines. Fuel cells also generate about the same waste heat as gas turbines; this heat can (and should) be used in a cogeneration setting whenever possible.

The George W. Bush administration made a big push to commercialize fuel cells, placing them at the center of an initiative to create a fuel cell vehicle known as the FreedomCAR. However, fuel cells have remained stubbornly expensive, far above the costs of CHP and wind and roughly equal to PV. Fuel cells large enough to power a building are good sources of backup power, in place of much noisier and polluting diesel generators. Overall, however, fuel cells are likely to remain a specialized form of DG used in remote locations where gas is available or as a medium-scale backup power source.[71]

Choosing among the Options

Table 8-1 summarizes the results of our supply technology tour. The second column of the table shows the representative *current* costs of power for each option as observed by regulators or measured by markets, and the third column does the same for projected costs as of 2030. In the fourth column, I add in the costs of carbon emissions based on an arbitrarily selected emissions price of $50/ton CO_2.[72]

The cost figures are a representative snapshot, and they have some important limitations. As noted many times already, renewable resources in particular are highly site-dependent, and costs may vary over a wide range from site to site. State and local policies may greatly affect the relative costs of two options, as when the California Solar Initiative gave rebates directly to customers installing PV systems.

Cost estimates for 2030 have even larger limitations. These figures reflect technical progress, which lowers the capital costs of every technology, some more than others. However, we cannot forecast how support policies will change, even though these policies could easily change the economic ranking of different power options. Finally, the 2030 estimates don't take into account resource depletion for any technology types—fossil fuel prices are constant in inflation-adjusted terms and the renewable resources are all at the same approximate quality. In the next chapter, we'll eliminate this assumption and discover that the size and pace of development can deplete the cheapest resource supplies and force costs up, especially in periods of rapid change.

The fifth, sixth, and seventh columns of the table contain further information on costs, constraints, and uncertainties. The fifth column notes the main drivers of uncertainties in observed costs. The sixth column lists the main factors affecting the systemic ("indirect") costs or benefits related to each power source, such as the savings in transmission and distribution investment from DG or the costs of integrating variable sources. The seventh and final

Table 8-1. Supply Technologies: Average Costs and Major Issues (All Costs Adjusted for Inflation)

(1)	(2) Observed Cost Today**, Including Tax Credits (¢/kWh)	(3) Observed Cost 2030 (¢/kWh), Excluding Current Tax Credits (no carbon cost)	(4) Observed Cost 2030 (¢/kWh), Including Current Tax Credits ($50/Ton cost CO₂)	(5) Cost Depends Strongly on . . .	(6) Indirect Utility-Incurred Costs or Benefits	(7) Other Issues
			Large-Scale Sources			
(1) Natural Gas Combined Cycle*	4.3–7.8	4.2–7.7	6.2–7.7*	Price of natural gas Price of carbon emissions	Over reliance on gas-reliability and price shocks	Supplies seem ample, but concerns over hydro-fracturing
(2) Coal with Carbon Capture and Sequestration (CCS) #, ##	7.7–15	6.6–15	7.1–15	Price of coal Technical progress Tax credits and other subsidies	Cost of carbon emissions permits under cap-and-trade rules and other greenhouse gas limits Technical risks of CCS lead to higher financial and regulatory risks	Technology must prove costs-effective at scale New questions about size of resource base and environmental concerns over mining practices Critical need for sequestration regulatory framework Scale and timing of sequestration of infrastructure

Table 8-1. Continued

(1)	(2)	(3)	(4)	(5)	(6)	(7)
	Observed Cost Today**, Including Current Tax Credits (¢/kWh)	*Observed Cost 2030 (¢/kWh), Excluding Current Tax Credits (no carbon costs)*	*Observed Cost 2030 (¢/kWh), Including Current Tax Credits Including current cost CO$_2$*	*Cost Depends Strongly on . . .*	*Indirect Utility-Incurred Costs or Benefits*	*Other Issues*
(3) Nuclear Power#	5.7–13	5.7–13	5.7–13	Cost of construction Loan guarantees or other supports	High levels of financial and regulatory risk Waste disposal	Challenges to Supply chain
(4) Onshore Wind	4.6	5.1	5.1	Quality of wind at each site Wind turbine prices Tax credits and other supports	Output variability requires investments in balancing resources and new operating procedures Costs of transmission expansion	Regulatory approvals for transmission Technical limits on amounts of wind that can be integrated into the grid
(5) Offshore Wind	8.0	8.0	8.0	Quality of wind at each site Wind turbine prices and new operating procedures	Output variability requires investments in balancing resources and new operating procedures	Same as onshore wind

(6) Photovoltaic Solar (Large-Scale Plant)	14.6	13.2	13.2	Overall construction costs	Tax credits and other supports	Costs and efficiency of PV cells	Tax credits and other supports	Output variability requires investments in balancing resources and new operating procedures	Transmission is critical	Cost of backup/storage integration; inherent system limits
(7) Concentrating Solar Power Plants	15.2–26.6	14.4	14.4	Construction costs and plant performance	Tax credits and other supports	Cost of transmission to sunny areas	Cost of balancing system loads with plant production	Requires large land area	Transmission is critical	Technology needs verification	Water use can be an issue	In-plant storage is valuable but as yet unproven

Table 8-1. Continued

(1)	(2)	(3)	(4)	(5)	(6)	(7)
	*Observed Cost Today**, Including Current Tax Credits (¢/kWh)*	*Observed Cost 2030 (¢/kWh), Excluding Current Tax Credits (no carbon costs)*	*Observed Cost 2030 (¢/kWh), Including Current Tax Credits ($50/Ton cost CO_2)*	*Cost Depends Strongly on . . .*	*Indirect Utility-Incurred Costs or Benefits*	*Other Issues*
(8) Biomass Power Plants	5.6	5.6	6.1	Cost of fuel, which is often site specific	Tax credits and other supports	Energy crops must be grown sustainably and not compete with food production
(9) Geothermal Power Plants	5.4	6.6	6.6	Quality of natural heat resource	Tax credits and other supports	Limited to high-potential sites in western United States
Small-Scale Sources						
(1) Natural Gas Combined Heat and Power©	3.9	3.4	6.1	Price of natural gas, site-specific conditions, and state regulations / Tax credits	Reduced utility control can lead to higher system balancing costs and backup power requirements	Constrained by practical difficulties, land use patterns, and some utility policies

Technology						
(2) Natural Gas Microturbine	9.2	7.3	10.9	Price of natural gas Technical progress	Same as above	Avoided upstream generation, transmission, and distribution investments; may be included in observed cost	Unfamiliar technology with more complex installation
(3) Natural Gas Fuel Cell	19.1	13.7	13.7	Tax Credits Technical progress Price of natural gas			
(4) Photovoltaic Solar (on Rooftop)	33.9	32.9	3.29	Costs and efficiency of PV cells Site-specific installation costs Tax credits and other supports	Output variability requires investments in balancing resources and new operating procedures Avoided upstream generation, transmission, and distribution investments; may be included in observed cost	Government and utility incentives play a critical role in making systems affordable

Table 8-1. Continued

(1)	(2) Observed Cost Today**, Including Tax Credits (¢/kWh)	(3) Observed Cost 2030 (¢/kWh), Excluding Current Tax Credits (no carbon costs)	(4) Observed Cost 2030 (¢/kWh), Including Current Tax Credits ($50/Ton cost CO_2)	(5) Cost Depends Strongly on . . .	(6) Indirect Utility-Incurred Costs or Benefits	(7) Other Issues
(5) Small-Scale Wind Turbines	20.2	13.7	13.7	Quality of wind at each site	Output variability requires investments in balancing resources and new operating procedures	Government and utility incentives play a critical role in making systems affordable
				Wind turbine prices		
				Tax credits and other supports	Avoided upstream generation, transmission, and distribution investments; may be included in observed cost	

Notes:

* Lower price based on $4/MMBtu gas; upper price $10/MMBtu; see Appendix 2.

Coal prices assumed constant at $21/MWh (2008 dollars).

** Costs do not reflect renewable energy credits.

@ Reflects a 50% reduction in costs as a credit for value of heat produced.

Upper end reflects the range in American's Energy Future, National Research Council, 2009, and is *not* adjusted for technical progress by 2030.

Sources: Author's calculations; see Appendix 2.

column lists some overarching issues and constraints applicable to each generator type.

The first rows in the table show the large-scale sources, followed by a second group of rows for the four main types of DG. In comparing these two sections it becomes obvious that the observed costs of small-scale generators are still much more expensive than their large-scale counterparts, with the exception of industrial cogeneration, which is one of the cheapest power sources around. The third column reveals that this gap does not decline over time, largely because both large- and small-scale technologies improve the performance at roughly the same rates. Only at the top end of the ranges for nuclear and coal CCS do levelized costs compare, and even then the renewable figures do not include the costs of backup power.

We know that the true cost gap is smaller than these figures show because small-scale sources reduce the need for upstream generation, transmission, and distribution investment. We also know that these avoided costs are hard to measure, require extensive regulatory involvement, and are very site and system dependent. The gap in observable costs highlights the importance of the policies and market structure changes that allow these costs to be measured and reflected in utility supply decisions. As we saw in Chapter 5, the regulatory challenges involved in setting the prices for the value streams DG provides are extraordinary. Even with common federal policies and standards, most of the calculations themselves will have to be made and debated on a system-by-system and state-by-state basis.

The rows of the table begin with one devoted to natural gas combined cycle plants. The two numbers shown in the cost column represent costs at very low and very high gas prices, $4 and $10 per million Btu, respectively. The small-scale gas options later in the table follow suit. Among the large-scale sources natural gas remains the cheapest as long as prices stay around $6 or less and fuel remains plentiful; CO_2 costs are not that large a threat. However, it is dangerous to rely too heavily on gas, and many regions already have large amounts of gas-fired capacity. As to coal, the second row reveals that, even with the estimated costs of sequestration added, low-carbon coal plants remain below the costs of most forms of renewable energy. However, these numbers assume that CCS technologies work as promised and coal is plentiful, and these are two big ifs.

Nuclear plants have a very wide construction cost range, as shown in the third row. Depending on these costs, they could be competitive with all other forms of generation or more expensive than everything but solar power. This range brings with it large financial and regulatory risks. Due to the size and cost of building a single nuclear plant—well over $10 billion—it is common to hear a nuclear construction decision referred to as a "bet the company strategy."

Among renewable sources, large-scale onshore wind (fourth row) is close to being cost-competitive now, depending on the quality of the specific location and grid integration costs. Biomass and geothermal power (eighth and ninth rows) are also in line with the costs of coal and gas power, but this is the result of a bit of circularity: utilities generally buy from or build on geothermal or biomass sites wherever and whenever the quality of the nearby resource is such that the costs of power match what they pay for a natural gas plant. The issue with these technologies is that the supply of sites with costs this low is highly constrained.

Solar PV remains the most expensive option, but its price (after tax credits) closes in on the other options and could well overtake them with technical breakthroughs. It is also the most versatile and modular of the supply options, and it is very popular with consumers and policymakers. Concentrating solar power, strictly a large-scale option, is a bit pricey now, but will be valuable if it can incorporate storage, continue its rapid progress, and get transmission.

Comparing the third and fourth columns, you may be surprised to see that putting a price on carbon does not knock the fossil fuel technologies out of economic contention, at least a price of $50 per ton of CO_2. Of course, CO_2 prices could be much higher than this or they could be lower, but as long as we have the capital and fuel cost numbers for CCS coal and gas combined cycle plants close to right, a carbon cap-and-trade policy seems far from a death knell for traditional fuels employing carbon removal technologies.

In summary, we know that there are many supply options with low or no greenhouse gases that are likely to work, or work already, but there are still a lot of questions about performance and cost. The commodity energy from small-scale options is still more expensive than large-scale sources, rendering regulatory policies that price the other DG value streams very important. Nearly every option is expected to get cheaper over the next twenty years, but the guesses—and that's what they are—show a lot of variation.

Each new type of resource also has its own set of uncertainties and risks—some unique, some shared. Natural gas can be pricey. Carbon capture and sequestration technology isn't available yet, there is no regulatory system for sequestration, and there are concerns over the supply of coal and the impacts of coal mining. The next generation of nuclear plants have uncertain capital costs and many supply chain limitations. Onshore wind is inexpensive, but requires substantial new transmission supplies and backup power and is presently limited due to integration issues. Photovoltaic, solar thermal, geothermal, biomass, and hydrokinetic each have their own challenges.

Amidst all of these questions and uncertainties utilities have to keep the lights on. They know their rate of sales growth is highly uncertain. They know the amount of DG on their systems will grow, depending greatly on their cost levels and local and national policy supports. With or without increasing sales, they know new plants will be needed to replace older units being retired, and that greenhouse gas limits will force many high-carbon plants into early retirement.

Deciding how much new supply is needed and choosing the generators that fill the need is generally the biggest single decision a utility CEO will ever make. Thousands of jobs, tens of billions of dollars, and the economic and environmental future of the utility's region is often at stake. The decision needs buy-in from many state and local officials, including state regulators and siting boards, as well as the business community, labor leaders, environmentalists, and the utility's own rank and file. And though it is extremely common to make small adjustments in the plan over time, big mistakes are fatal.

If you're wondering how utilities cope with this mess, meet Jim Jura.

CHAPTER NINE

Billion Dollar Bets

JAMES J. JURA, CEO of the Associated Electric Cooperative in Springfield, Missouri, is no newcomer to the power business. Before coming to Associated, he served as the top administrator of the Bonneville Power Administration, the federal government's mammoth system of dams and transmission lines in the Pacific Northwest. With an earlier stint at the White House Office of Management and Budget, he's no stranger to finance and accounting, either.

But as head of an electric cooperative that still projects sales growth, he faces most of the challenges typical of utilities today. Associated is a kind of super-cooperative, with 850,000 customers, eight power plants, and a transmission system that reaches to four states. It is governed by six regional co-ops which are in turn owned by fifty-one distribution cooperatives in Iowa, Oklahoma, and Missouri. In effect, it is the exclusive, self-owned power supply arm of these fifty-one retail distributors.

Jura reports to a twelve-member board elected by the six regional co-ops. Each year, he brings them a proposed long-term plan for their approval. "I work for a co-op created to do a very simple thing," Jura says, "provide a reliable power supply at the lowest possible cost." Yet "there's a tremendous amount of

uncertainty out there." Jura is proud of the fact that Associated was the first utility in the state to purchase wind power on a large scale. Despite his energy efficiency programs, he needs more supply in the next two years. He knows the
Smart Grid and distributed generation (DG) are coming, but he doesn't see
them as large enough, fast enough, or cheap enough to close his resource gap.

Under his direction, Jura's planners perform an annual ritual of examining
every one of their supply options, from another coal-fired plant (beyond the
two they already own), to wind, solar, natural gas, and a share in a new nuclear
plant developed by a nearby investor-owned company. Here's how they do it.

Planning for the Next Few Decades

The first step in a supply plan is to determine the need for new supplies. In today's climate, this comes from replacing plants that become too old or expensive
to operate, plus a possible sliver of new sales growth remaining after demand response (DR), energy efficiency programs, and customer-generated DG are sub-
tracted. Because not all plants supply the same type of power—some are better
for round-the-clock supply while others are better for episodic service—the
timing and nature of the need must be characterized.

Planners next create a handful of supply portfolios that fill the need and use
computer models to estimate the costs of each portfolio over the next twenty to
thirty years. Finally, they ask themselves what couldn't be measured well in their
computer-generated cost estimates, including such things as the systemic cost
savings from increased DR and DG.

Candidate portfolios are always bedeviled by uncertainty over the options
we just examined. Should a utility build a coal plant in the next five years assuming that carbon capture and sequestration (CCS) technology can be added on in
the decade following? Will natural gas prices be high, low, or medium a decade
from now? As we saw in the last chapter, for relatively untested large-scale technologies the cost of the plant itself is another major uncertainty. Will there be
breakthroughs that scramble the relative cost of the options?

The life cycle cost of these options is estimated using computer models that
actually simulate the operation of the system hour by hour for up to thirty years.
To account for a wide range of uncertainties, many scenarios are run. As an ex-

ample, every portfolio's cost would be estimated with several price forecasts for natural gas as well as different scenarios involving sales growth, future plant costs, and so on.

Odd as it may seem, the advent of a carbon cap-and-trade system does not in itself pose any momentous new hurdles to this planning process. Power plants are already subject to many types of emissions rules, and the ability to factor carbon emissions into a supply plan is already built in to most utility planning software. Many planners started including an estimated cost for carbon allowances years ago.

This doesn't mean, however, that climate change uncertainties are small or easily managed. The predicted prices of carbon emissions permits vary tremendously. Old coal-fired power plants produce about 1 ton of CO_2 for every megawatt-hour (MWh). Without carbon prices, 1 MWh of coal power costs about $30. Estimates of the price of 1 ton of CO_2 vary between $5 and $500 in various simulations of carbon markets through 2050. Carbon prices at these extremes add anywhere between 17% ($5 added to $30) and 1666% ($500 added to $30) to the original price of coal power. And it isn't simply a matter of guessing where prices will land within this range, as carbon allowances will be a traded commodity likely to have considerable price volatility.[1] In Europe, for example, carbon allowance prices have varied between almost nothing and $34/ton since 2005. So while we know *how* to factor uncertain carbon emissions prices into utility plans, the fact that we don't know allowance prices adds a whopping new source of uncertainty to portfolio costs estimates.[2]

After months of calculation, consultations with stakeholders, and internal debate, the moment of truth arrives and a portfolio is selected. Cheapest doesn't always win in these contests. In the first place, different portfolios will have different periods of being cheaper and more expensive over the decades, and the answer to "what's cheapest?" depends heavily on your time horizon. And apart from the numbers showing the option with lowest expected cost, the size and allocation of risks, political considerations, and administrative feasibility play an important and sometimes determinative role.

What about Jim Jura and Associated? After looking at the options, the board concluded that nuclear was just too expensive and risky. Although coal was still viewed favorably as a long-term option, Jura called it "pretty much over for now until we get more policy certainty." Instead, Associated chose a combination of a new wind farm and two new natural gas plants, one combined cycle plant, and one peaking unit. Noting the uncertainties about fuel and carbon prices, Jura called wind "a hedge for the future." But wind wasn't an option if deployed by itself. "Out here, on a summer day the wind blows a lot at 1 a.m., but it tapers down pretty fast from there."

Associated was fortunate in that it is located in a region where both natural gas and wind are readily available. Had it been in New England, for example, state authorities may have been concerned about overreliance on natural gas—New England is at the far end of the U.S. gas pipeline system. In the U.S. Southeast, it would probably have needed to consider biomass power, as neither wind nor solar are especially strong there, or it may have chosen to gamble on nuclear or coal.

If he had been in a region where state policymakers or his customers were demanding more DG, he would have had to determine how much would be installed, the integration impacts on his system, and how much he would charge for backup services. He might have chosen to add some local storage in a few parts of his system where there were an unusually large number of small generators or possibly configured some distribution or subtransmission lines differently.

Associated adopted a strategy common to many utilities with access to cheap wind these days—combine it with natural gas plants and some energy efficiency. However, as Jura's comments suggest, this strategy is something of a stopgap until there are more options, more policy certainty, and more progress on the Smart Grid, DR, and DG.

Handicapping the Supply Scenarios

Across the longer horizon, there are a handful of ways the industry's supply mix might evolve—call them scenarios. Formally, scenarios are highly distilled views of where the future will end up, useful because they help guide what the legendary planner Pierre Wack called "the gentle art of reperceiving."[3] The costs and uncertainties we saw in Table 8-1 suggest that any of these supply scenarios remain possible, though some face larger challenges than others.

Scenarios make it easier to spot what planners call *signposts*—future events that portend a shift in the direction toward one scenario or another. A signpost may be a change in the availability of a key raw material, a technological breakthrough, or a political development.

The first scenario, the one belonging to Sheila the blogger (whom you'll remember from Chapter 7) is *Small Scale Wins*. Here the combined effects of DR and energy efficiency, coupled with the greater use of small-scale and downstream storage, eliminate the need for added upstream supply.

In the opposing vision, the *Traditional Triumphs*. Here, coal with CCS and/or new nuclear plants become cost-effective sources for baseload power, just as they are in the world's current system. Other planners might call this scenario business as usual, except that in this case business as usual includes a price on carbon emissions.

If all power can't be made at the local level, perhaps it can nonetheless be 100% renewable. *Completely Green* aggregates large- and small-scale renewable sources, along with DR and energy efficiency, in a scenario that relies on no new gas, coal, or nuclear power. Vice President Gore's Repower America campaign called for this outcome in late 2008; more recently, Google's energy team and other researchers have produced more detailed proposals along these lines.[4] We have to ask ourselves, what if none of these scenarios occurs? I call this outcome *Most of the Above*. In this final catchall scenario, no one type of supply dominates, and the power grid evolves into an even more diverse and distributed fleet of generators than it is today—nearly all of them low or no carbon.

Let's start with *Small Scale Wins*, the scenario in which all new large-scale plants and lines are no longer needed. In spite of the hype surrounding the Smart Grid, the higher costs and regulatory impediments to rapid adoption make this scenario exceedingly unlikely.

As Table 8-1 revealed, other than combined heat and power (cogeneration) the observable costs of DG are still two to three times as high as large-scale sources, and neither technological change nor a price on carbon will close the gap. Regulatory and other support policies will also help, but they face many challenges and implementation time lags.

If there is an economical path to this scenario it undoubtedly involves a massive push to expand the use of combined heat-and-power (CHP) plants.

There are somewhere between 76,000 and 85,000 MW of this type of plant in use today at about 3,300 sites.[5] Several analysts have estimated that a concerted effort to encourage CHP plant use could lead to additional installations of 80,000 to 100,000 MW by 2030.[6] There is every reason to believe that this cost-effective potential is out there; the questions are whether industrial plants and large commercial building developers will make this practice routine, and whether local distribution utilities can and will expand their systems fast enough to absorb this much local generation.[7]

It is less likely, but remotely possible, that an equally large increment of new capacity would come from solar photovoltaic (PV) panels. This would require that their costs drop much faster than the 23% drop predicted by 2030 that the Energy Information Administration (EIA) estimates, along with an aggressive rollout of the Smart Grid, local storage, federal and state tax supports, and a supportive regulatory climate. All the necessary ingredients would have to align with nearly magical precision to back out all other sources by 2030.[8]

Another factor that adds weight to my doubts about this scenario is the rather downcast assessment of DG recently released by the State of California. No U.S. state has a more favorable posture toward DG, or has spent more to improve utility policies toward this technology, than the Golden State. Nonetheless, the California Energy Commission recently concluded the following:

> The DG industry is still a nascent industry that survives despite difficult market conditions. Many projects are highly customized and rely on incentives. The industry is fragmented with many "small" developers installing PV and natural gas engines provided by large, well-established equipment suppliers. There is fragmentation by technology type and diverse business models. . . . Due to low penetration rates, DG installations do not have a large impact on, nor are they integrated with, the state's electric and natural gas infrastructures.[9]

The Commission estimated that a full-court press toward all forms of DG could produce 26% of the state's peak electricity demand by 2020, or a total of 18,600 MW. Of this total, however, 60% would have to come from CHP plants larger than 20 MW. Out of the remaining 7,610 MW, 3,000 MW would come from PV,

3,600 MW would come from small CHP (including biomass), and 10 MW would come from small wind.[10]

Looking beyond 2030 I think it is much more likely that smaller-scale supply sources will become the norm. By this time a substantial portion of the distribution network will have the features needed to integrate DG. It is also likely that this is the time frame in which natural gas and coal will become further depleted and more expensive in the United States, if it hasn't happened already. Finally, we will know by 2030 whether the next generation of nuclear and CCS coal plants are going to work, what they're going to cost, and whether the regulatory and financial risks will allow investments of this magnitude and lead time.

If the current plans to expand the grid fail, it is also possible that a dynamic could develop over the next twenty years that accelerates DG's growth. We know that state portfolio standards mandate roughly 208,000 MW of new renewable capacity by 2030, and that a federal renewable mandate could add to this total. If transmission expansion makes it impossible to add this much large-scale renewables, policymakers will work harder to promote and finance small-scale renewables to meet renewable mandates. Meanwhile, the grid will become even more overloaded as upstream renewables and their backup sources crowd onto the limited capacity, a bit like we see today. Further strains on the large-scale grid could create price spikes and greater reliability fears, prompting more customers to install DG and storage to protect against these two phenomena. As planners and stakeholders see a trend toward DG they may be even more reluctant to add transmission, exacerbating the upstream problems that gave DG a boost in the first place. Figure 9-1 depicts this possible, though not entirely likely, sequence of events.

The unlikely success of *Small Scale Wins* should not be seen in any way as a prediction that Smart Grid technologies will fail. Quite the contrary. The scenario refers only to the main reliance on small-scale sources, not the new controls and functions the Smart Grid will enable. As Smart Grid capabilities become more widespread, the ease of incorporating DG and medium-scale resources will steadily increase. At the same time, large upstream plants will be around for a very long while. There is every reason to encourage all economical

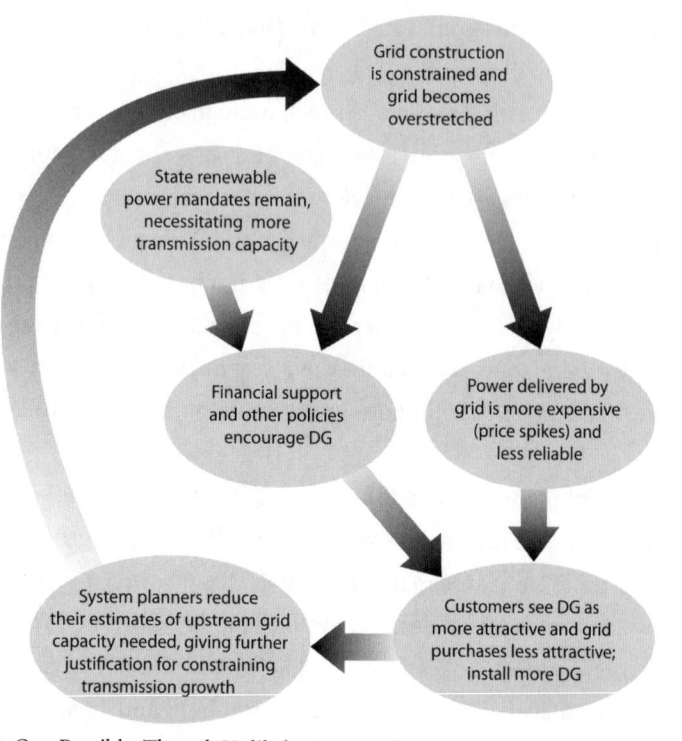

Figure 9-1. One Possible, Though Unlikely, Process That Would Accelerate Distributed Generation (DG).

DG, but the high costs, regulatory adjustments, changes in land use and construction practices, and time required for the Smart Grid to mature all suggest that DG cannot be our *exclusive* source of new low-carbon power in the current half-century.

Traditional Triumphs and *Completely Green* are in direct opposition; better prospects for one equate to a worse outlook for the other. At the intersection of these two scenarios, nearly everyone in the power industry expects gas-fired power to continue to play a very large role for at least the next decade. It is versatile, proven, cheap, and ideal for supplying the balancing power needed to smooth out wind and solar production. It is nearly impossible to imagine any future without natural gas as the near-term choice for backing up solar and wind, exactly as Jim Jura and Associated decided in their plan.

The prospects for low-carbon coal and nuclear, though much less certain, are strong enough to keep them in contention. Neither of these can yet be considered a proven commercial technology in the United States; their survival absolutely depends on constructing commercial units that perform well at competitive prices. In both cases, there is strong research and development (R&D) and early commercialization support, which will be essential to meet this goal. As noted earlier, the U.S. Department of Energy just awarded grants to several large CCS projects and is scheduled to give $16 billion in loan guarantees for an initial first group of new nuclear plants, along with production tax credits for power from the first 6,000 MW.[11] As to coal, it is politically unrealistic to expect that U.S. climate legislation will pass without significant support for coal plants with CCS. As an example, the Waxman–Markey climate legislation allocates bonus allowances of up to 72,000 MW of CCS capacity in two phases and allows coal-fired utilities to set up a unique self-funded R&D cooperative, in addition to a mandate that all future coal units install CCS after 2020.[12]

For both coal and nuclear technologies, R&D efforts are occurring on a worldwide basis. FutureGen has utilities with operations on six continents, and the European Union has pledged to build up to twelve coal plants with CCS to jumpstart the industry.[13] China also has an active CCS research program, and much nuclear R&D is also now offshore. The increasing use of these two fuels in Asia and Europe means that U.S. companies and electric customers will not be the sole public funding source for R&D, and will benefit from international spillover.

Beyond R&D and capital costs, coal and nuclear's success will also depend on signposts in the regulation of their fuel cycles. Sequestration must have a framework in place that enables financing and operation of a gigantic national CO_2 piping and storage infrastructure. Nuclear plants will need stable licensing conditions and waste storage. And, of course, fuel must continue to be available, with acceptable environmental costs, at prices that keep these forms of energy competitive.

Coal has been a leader in the sector for a long time, and the first generation of nuclear plants overcame many of these issues. Coal and uranium come from domestic or relatively stable and secure foreign suppliers, giving these fuels one very attractive feature compared, for example, to our oil supplies. Economical CCS and nuclear plants will unquestionably make it much easier for the world to reduce greenhouse gas emissions. In view of these policy pluses, the future of these sources hinges on their performance and costs.[14]

If nuclear and coal both stumble, *Completely Green* will be the inevitable beneficiary. While it might be tempting to view this as extremely likely, renewable energy faces many challenges of its own. In most forecasts PV remains expensive, even through 2030, and so does concentrating solar power (CSP). CSP also needs siting approvals and must manage its water demand. Offshore wind has high siting costs and risks. Biomass resources face competing land use demands and environmental questions. Above all, large-scale renewable development requires transmission resources, integration and backup investments, and Smart Grid deployment close to home.

There is little question that renewable energy supplies have the physical potential to satisfy all our energy needs, but this is not the same thing as saying that they can supply all our needs as cheaply as we can with another century of (tapering) fossil fuel use. Analysts who have looked carefully at the cost of producing renewable power from progressively lower quality sites find that these costs will rise as we use more of it. DOE's massive study of wind energy, for example, found that wind costs would approximately double, from around 6 cents/kWh to 12 cents/kWh, as U.S. wind capacity grew to 800,000 MW.[15] Another thorough assessment of all renewable energy options for the State of California found that the cost of renewables per kilowatt-hour would rise more than 500% as cumulative development reached 250 billion kWh/yr.[16]

As with nuclear power, renewable power technologies have their own supply chain limits, and pressing against these raises costs or is simply unfeasible. California's analysis of its ability to meet its ambitious renewable mandates reports that the time needed to develop projects and the relatively short period until 2020 requires that multiple options and contingency plans will be needed, raising the cost of achieving the goal if it can be achieved at all.[17] In an interview with one of the report's authors, he relayed that one renewable industry representative told him that they simply could not build more than a set amount of new capacity in a decade.[18]

And the Winner Is . . .

Provided you do the judging around 2030, or better yet closer to 2050, any of the four supply scenarios could come true. The most likely answer though, is *Most of the Above*. Although the directional shift toward renewable and smaller sources is unstoppable, the lowest-cost approach to meeting strong carbon limits is likely to include either or both coal and nuclear units, provided at least one works at its projected cost.

This finding is echoed by most researchers who have studied the future power systems. Observers as diverse as the U.S. Environmental Protection Agency, climate expert Joe Romm, the Electric Power Research Institute, Charles River Associates, and the Stanford University Energy Modeling Forum have all formulated broad national least-cost plans or simulations indicating a significant role for CCS coal, nuclear, and many types of renewables. A handful of exceptions, including Google's energy plan and one from the Natural Resources Defense Council (NRDC), leave out one of the two traditional sources (CCS or nuclear) but keep the rest.[19] Table 9-1 shows the distribution of projected supplies in 2030 from each main type of fuel from a sample of these studies.

Some experts go beyond least-cost prediction to say that, due to the urgency of emissions reductions and of the extent of current and projected coal and nuclear power, one or both of these two sources *must* be a part of the future mix. "There is no credible pathway towards prudent greenhouse gas stabilization targets without CO_2 emissions reductions from existing coal power plants," says MIT's Ernie Moniz, a former undersecretary of energy and professor of nuclear engineering (he thinks nuclear is essential too).[20] In an interview quoted in *Climate Wire*, Judi Greenwald of the Pew Center on Global Climate Change, a respected climate think tank, said it more simply: "either CCS or nuclear power has to work 'in a big way' or else the planet is in trouble."[21]

This modeling and research suggests additional reasons favoring *Most of the*

Table 9-1. Projected Electric Power Resource Mix in Five Assorted Studies (2030)

Percentage of Supply From . . .		Environmental Protection Agency	Google.org	National Renewable Energy Laboratory	Union of Concerned Scientists	Charles River Associates
Fossil Fuels	Coal with Coal Capture & Sequestration	5%	0%	0%	0%	3%
	Coal	46%	0%	33%	6%	38%
	Natural Gas & Oil	13%	24%	28%	58%	16%
	Subtotal	64%	24%	61%	64%	57%
Nuclear	Nuclear Energy	20%	9%	10%	12%	27%
	Subtotal	20%	9%	10%	12%	27%
Renewables	Hydro	7%	6%	8%	4%	7%
	Wind		31%	12%		
	Solar	9%	21%	5%	21%	9%
	Other Renewables & Distributed Generation		8%	5%		
	Subtotal	16%	67%	30%	24%	16%
Total		100%	100%	100%	100%	100%

Note: Numbers may not total to one hundred due to rounding.

Sources:

"EPA Analysis of the American Clean Energy and Security Act of 2009, H.R. 2454 in the 111th Congress," U.S. Environmental Projection Agency, June 23, 2009. Available at: http://www.epa.gov/climatechange/economics/pdfs/HR2454_Analysis.pdf

Jeffrey Greenblatt, "Clean Energy 2030: Google's Proposal for Reducing U.S. Dependence on Fossil Fuels," Google.com, October 6, 2008. Available at: http://knol.google.com/k/jeffery-greenblatt/clean-energy-2030/15x31uzlqeo5n/1

Patrick Sullivan, Jeffrey Logan, Lori Bird, and Walter Short, "Comparative Analysis of Three Proposed Renewable Electricity Standards," National Renewable Energy Lab, May 2009. Available at: http://www.nrel.gov/docs/fy09osti/45877.pdf

Rachel Cleetus, Steven Clemmer, and David Friedman, "Climate 2030: A National Blueprint for a Clean Energy Economy," Union of Concerned Scientists, May 19, 2009. Available at: http://www.ucsusa.org/global_warming/solutions/big_picture_solutions/climate-2030-blueprint.html

Scott J. Bloomberg and Anne E. Smith, "Analysis of H.R. 2454 (the Waxman-Markey)," Charles River Associates, June 17, 2009. Available at: http://www.nma.org/pdf/040808_crai_presentation.pdf

Above that are not captured by most current analyses. First, the models used to simulate climate strategies tend to have little or no ability to capture the difficulty or timing of new transmission construction, which is a significant constraint on renewables. We also know comparatively little about the steepness of the supply curve for some renewables, including whether integration costs will increase or decrease as more renewables are added and technology improves.

There is also a value in diversifying fuel supplies supported by considerations of both price stability and national security. This is the same portfolio value theory used to build sound investment portfolios; many utilities and regulators apply it in either a simplified or advanced form already. As power expert Fereidoon Sioshansi put it after a long study of supply options;

> [A]s I studied the various options to de-carbonize electricity generation even if one were to include developing countries, I realized the limitations of each technology, and it began to dawn on me that the task at hand is more daunting than I had imagined. Pushing too hard or too fast on one technology or option would merely result in problems somewhere else in the interconnected system.[22]

Finally, there is fairly strong political momentum behind all of the major forms of energy, including the major renewables, in the United States today. They have good trade associations and represent many American jobs scattered across many congressional districts. Each can be expected to fight hard to keep its share of the future supply market by way of R&D support, early commercialization subsidies, and many other types of regulations and mandates.

As the industry reaches for 2050, a continuation in the shift toward renewable energy and smaller sources is inevitable. The Smart Grid will become the norm, and downstream generation and storage will become far more common. Yet it is also nearly certain that we will expand the high-voltage grid and build thousands of megawatts of large power plants of some persuasion. The power industry is going to have to do it all over the next fifty years: engineer a complex, paradigm-busting distribution and information system, build transmission faster than ever in recent history, learn how to integrate vast quantities of variable renewables and small sources, and build new large plants from technologies that are mainly unproven. The estimated costs of the effort just until 2030 lie somewhere around $2 trillion.[23]

Who is going to do all this? Regulated utilities? Public power? Deregulated generation companies? Will these challenges force companies to specialize in one part of the industry or to integrate vertically into all parts of the supply chain? Above all, does governing this industry properly through its period of upheaval call for changes in regulation and industry structure? It is time we turned from the industry's amazing physical transformation to ask how we make a business out of all of this change, the subject of the next and final part of this book.

PART THREE

Business Models for the New Utility Industry

CHAPTER TEN

Energy Efficiency: The Buck Stops Where?

I F ENERGY EFFICIENCY (EE) is often the lowest-cost option, why don't we take more advantage of it? This is often posed as one of the great conundrums of American energy policy, sometimes to try to discredit the existence of low-cost efficiency or the idea that we have any successful ways of accessing it. The real answer lies in understanding the gap between what we measure as EE's cost and the economic actions of real-world consumers and energy firms. And the first step in closing that gap is in setting the right market conditions for those consumers and firms through policy mechanisms.

When we say an EE measure costs less than a new supply alternative, here is what we mean: if you spend the capital to buy and install more efficient technologies in a building, the added cost will more than come back to you in savings over the lifetimes of the technologies. If you buy a more expensive air conditioner that's more efficient, your electricity bill savings over the next five or ten years are larger than the added cost of buying the better unit.

Cost-effectiveness conclusions like this come from comparing the *net present value* of the costs of EE to the alternative of building and fueling a new power plant over its lifetime. A net present value takes a stream of costs and benefits stretched out over years and reduces the stream to an equivalent single dollar value paid today.

Investing in EE and buying more power are very different options, especially in the timing of the money outlays and the return of benefits. You have to pay 100% of the costs of buying a more efficient appliance before you get a single unit of benefit; the benefits are then stretched out over the next several years. In contrast, regardless of how much it cost to build the power plant we buy from, we only have to pay for it 1 kilowatt-hour at a time.

The use of net present value ensures that any comparison of the two options is accurate from the standpoint of modern financial theory. The problem is people don't act based on net present value. There are dozens of ways each of us could save money over the next several decades if we spent money now and waited patiently for the savings to materialize. We might be better off in the long run, and society might be better off, but still we fail to act.

The notion that people do not act to maximize their welfare, defined and measured by economic concepts such as net present value, is now a well-accepted field of study known as behavioral economics.[1] Thaler and Sunstein's popular book *Nudge* explains nicely why humans don't always act with "economic rationality":

> Those who reject paternalism often claim that human beings do a terrific job of making choices, and if not terrific, certainly better than anyone else would do (especially if that someone else works for the government). Whether or not they have ever studied economics, many people seem at least implicitly committed to the idea of homo economicus, or economic man—the notion that each of us thinks and chooses unfailingly well, and thus fits within the textbook picture of human beings offered by economists.
>
> If you look at economics textbooks, you will learn that homo economicus can think like Albert Einstein, store as much memory as IBM's Big Blue, and exercise the willpower of Mahatma Gandhi. Really. But the folks that we know are not like that. Real people have trouble with long division if they don't have a calculator, sometimes forget their spouse's birthday and have a hangover on New Year's Day. They are not homo economicus; they are homo sapien.[2]

The rationale for EE policies is strongly related to these behavioral insights. Real-world people can't absorb the information nor do the calculations needed to evaluate energy-savings opportunities, and they respond to complex choices with all-too-human inertia. In addition to pointing out that real people have trouble with long division, Thaler and Sunstein note that decision making is especially "non-economic" when people are "inexperienced or poorly informed and in which feedback is slow or infrequent"—an apt characterization of EE decisions.[3]

If policies can help energy users overcome these limits without offsetting waste or welfare losses, both society and the power customers who save will be better off. Part of the welfare improvement is in the form of lower life cycle power bills for those who conserve with no loss of comfort; the rest comes from lower costs, higher employment, and lower environmental impacts for everyone else.

Efficiency in National Climate Strategies

All over the world, nearly every study of climate policy options concludes that the single cheapest option for CO_2 savings is energy conservation. Among all conservation options, electricity savings are almost always cheapest, and they are excellent for boosting employment as well.[4]

Following a major assessment, the Intergovernmental Panel on Climate Change (IPCC) concluded that most countries of the world could reduce carbon emissions by 11 to 85% in 2020 by making buildings more energy efficient at less than $25/ton of CO_2 saved.[5] Nearly all other options cost more or have lower potential, including every type of low-carbon energy supply and most savings options in the transport sector.

These findings occur in essentially every other climate policy study, including those by McKinsey, EPRI, the Union of Concerned Scientists, EIA, the International Energy Agency, and the U.S. Environmental Protection Agency.[6] Climate policy experts across the political spectrum urge that energy efficiency be the first action taken to reduce greenhouse gases. Eileen Claussen, head of the Pew Center on Global Climate Change, calls energy efficiency "about as close as you can get to a silver bullet."[7] It is telling that the National Action Plan for Energy Efficiency, the most sweeping set of efficiency policy proposals in decades, was produced in 2006 by the Climate Change division of the EPA along with the DOE.

For all the climate policy benefits of EE, however, you haven't accomplished anything until you get someone with sufficient information, motivation, and capital to make an investment in greater efficiency. The industry has been confronting this challenge for decades with very mixed results. Improving this performance is essential for a successful climate policy.

The Barriers to Greater Efficiency

Policy wonks often use the word *barrier* to refer to all of the factors that cause consumers not to install EE measures that pay for themselves over the measures' lifetime. The barriers have been thoroughly studied and documented.[8] The main barriers—information, capital availability, transactions costs, and inaccurate prices—can be described briefly as follows:

Information. EE is a field of expertise and specialization just like energy supply. It takes training and experience to keep up on the cost and performance features of energy use technologies and evaluate savings options for a specific building or application. We should not expect most homeowners or businesspersons to have the knowledge to do this. Furthermore, EE options raise important decisions and tradeoffs right inside one's home, office, or factory—tradeoffs much more complicated than simply leaving your building alone and buying more power. Because so many of our structures and energy needs are unique, each energy-saving application must be somewhat tailored to each customer.

Researchers rarely measure or count as a cost the time required to educate consumers as to their options, or to find them expert assistance, but these costs can be large. As a result, EE programs that prescreen efficiency options for consumers and give them hands-on assistance choosing measures that work for their specific needs have proven to be valuable.

Capital availability. Every entity has a cost of borrowing money, whether from a bank loan, new equity, or rich Uncle Harry. As borrowers approach their credit limits, their borrowing costs rapidly reach a point where investments simply can't be financed. This is particularly harmful to EE because 100% of the cost of efficiency is paid before any savings are realized; it is by definition an all-capital option.

Even where businesses or consumers can borrow enough to finance efficiency they often hesitate to use up their borrowing power for this purpose. EE measures can take years to repay themselves. If circumstances change during that period—the home or business relocates, energy prices change, or other changes occur—that borrowing capacity might be needed for something else. When asked why they did not adopt an EE measure that will repay itself within a few years, many a businessperson has said roughly this, "I am in the business of making and selling (fill in the blank). That's what my investors invest in and that's what I am good at. I am not in the business of saving energy. Why should I devote my scarce capital to EE investments, which I don't understand and can't measure, when I am more confident I can make profits using my capital to do something I know I am good at?"

Transaction costs. Transactions costs are a technical name for what might be better called the hassle factor. Unlike buying more supply, which involves minimal intrusion into our daily routines, EE measures require construction and/or operational changes in homes or businesses. One has to be prepared to deal with contractors, engineers, occupancy delays, and so on.

Anyone who has ever managed a construction project knows that all sorts of things can go on, delaying or damaging equipment or property. One of my consulting partners once tried to convince a Las Vegas casino owner to install EE measures in a new building under construction. He and the local utility had arranged it so that the EE measures would be added as part of the overall construction project, but the measures would extend the construction schedule by five days. In exchange for the five-day delay, the casino owner would recoup the efficiency investment within one year and save millions on the power bill for years afterward. But the casino owner said no—fearing that it interfered with the carefully planned construction and opening schedule; even this minimal delay was not acceptable.

Inaccurate price signals. Many energy decisions are made by a builder or landlord who will not pay electric bills for the equipment or building at hand. Two-thirds of all water heaters, half of all furnaces, and a quarter to a third of all refrigerators are purchased this way.[9] In this case there is no direct payback to the builder or landlord who pays more for a more efficient technology. Even where there is no problem of this sort electric prices almost always understate the true cost of providing electricity. Electric regulators have generally resisted setting prices that reflect either the true cost of production or the unpriced externalities of power use. Higher, more accurate prices naturally encourage more conservation by increasing the value of the savings in the calculations we have been talking about. This is why dynamic pricing, discussed in Chapters 4 and 5, is so important.

Using the term *barriers* to discuss these four factors conjures up visions of some artificial boundaries that policymakers have mistakenly erected to prevent the EE we would naturally adopt otherwise. This isn't quite right. Apart from lousy price signals, policymakers have done nothing to *prevent* Americans from investing in EE—we just make our own private, self-interested choice to use what capital we have in other ways. The more accurate way to refer to barriers is to treat them as hidden, unmeasured costs that don't enter into economists' official figures. Along with behavioral inertia, these hidden and usually unmeasured costs account for the difference between the investments made by energy users and the investments that would be in society's best interest.

However we choose to view them, if we want to harvest more of our EE potential we need policies and business models that address the barriers directly, that is, intervene to lower hidden costs. When, for example, someone offers homeowners free energy audits by expert auditors, many homeowners take advantage of the offer. These auditors have the expertise needed and apply it to the individual homeowner's unique structure and preferences. Sometimes they provide installation advice or assistance as well. By doing all this, the hidden costs are reduced to practically nothing, and many more customers take advantage of efficiency measures that are cost-effective.

Getting the Prices Right

When the subject of government involvement in EE comes up, the first thing you're likely to hear is that "we can't possibly have a sound energy policy until we get energy prices right." Technically, the statement is accurate. However, like so much of the policy rhetoric in Washington, this statement sometimes carries with it some ideological baggage.

One aspect of getting prices right has to do with the time structure of electric rates. As we learned in Chapter 5, nearly all electricity prices in the country are set at an annual average rate level, often in quantity tiers (e.g., one price for the first 500 kWh per month, another price up to 1,000 kWh, and so on). It is much more efficient to adopt dynamic pricing and charge the approximate price of production each hour for generation. As we've seen, one of the most important features of the Smart Grid is that it automatically enables dynamic pricing.[10]

The second, very different aspect of getting prices right is the inclusion of externalities in electricity's price. Once again, there is no disputing the propriety or desirability of including externalities when it is feasible (often it isn't). Among energy production externalities, the 800-pound gorilla is the control of CO_2 emissions.

Getting prices right is on solid theoretical ground, and it is more or less doable. However, there is one seemingly tiny rhetorical leap from this position to one that rejects public energy efficiency programs. The rationale goes like this: Most electric regulatory policies have the effect of reducing electricity prices below "free market" levels. Since underpricing electricity causes people to use more of it, a combination of underpriced power plus publicly funded EE programs is a wasteful contradiction. It is like the government giving out free candy and then running programs for weight loss at the same time.

It is indeed regrettable that many utility regulations and pricing policies tend to suppress prices when they should be higher. However, suggesting that nothing should be done until the marketplace exhausts all the efficiency induced by deregulated prices is unrealistic, since deregulated prices are not a viable option in most of the United States today.[11] More importantly, regulated and public sector utilities are slowly getting better at sending accurate price signals. Finally, EE programs can sometimes be designed to correct for pricing deficiencies.

More importantly, an enormous body of evidence shows that the nonprice market barriers have a greater impact on EE than inaccurate prices. Raising prices to economically accurate levels certainly helps, but its impact is much smaller than the impact of making customized information or low-cost capital available, or helping out with the hassle factor.

A numerical example will show you why this makes sense. Suppose that adding the cost of carbon externalities to electricity increases electric prices by about 2 cents/kWh. When we plug the new numbers in a net present value calculation, we find that an efficiency measure will now pay itself back in five years rather than seven. This is good news, but there is ample evidence that consumers will not buy technologies with *either* five- or seven-year paybacks without assistance. Conversely, if you told these consumers you would give them a loan, with no repayment required on their part, and they could keep the savings after the loan was repaid from the energy savings, they will buy the technology in either case. A formal analysis of this phenomenon by economists Adam Jaffe, Richard Newell, and Robert Stavins found that upfront subsidies were three to eight times as effective as higher prices.[12]

Sometimes one finds that an argument that we should wait to implement EE until we have prices right really reflects a veiled discomfort with the degree of "social engineering" that EE programs reflect. EE programs try to get customers to change their energy choices or even bribe them to do so. Like most other government-induced policies, they redistribute economic development from power plant construction to investment in customer buildings and equipment.[13]

Ultimately, this is a question of political and social judgment. Governments guide and limit the choices of their citizens in dozens of ways when they judge that such limits leave their citizens better off. If EE policies demonstrate results at reasonable costs—a crucial requirement—there is now widespread agreement that government policies should enable information and capital that reduce energy use.

Energy Efficiency Policy Paths

Beyond sending good price signals, energy efficiency requires a raft of policies that start with strong energy R&D and end with programs that actually place efficient technologies in residential and commercial structures, where nearly all power is used. Although utilities often play an important role in R&D, the essential policy questions relate to its role at the *deployment* end of the spectrum.

At a high level, there are really only a handful of policy approaches that have established strong records of achieving substantial efficiency gains at costs lower than those of more supplies.[14] The first proven approach is building codes and efficiency standards for all types of household and commercial appliances. Codes and standards save enormous amounts of energy. Appliance efficiency standards have saved Americans more than $30 billion since the 1987 National Appliance Energy Conservation Act,[15] with projected total benefits reaching $400 billion by 2030,[16] not counting future standards that DOE is already required to implement.[17] Several states that have adopted strong building standards report that a modern code-compliant building uses 75% less power than the same-size building constructed before codes were adopted.[18]

But codes and standards also have their limits. They are exceedingly unpopular with the construction and real estate industries and appliance manufacturers; these powerful industries often oppose substantial increases in their mandates and succeed in weakening them. There are no national building codes, so each state or locality can adopt whatever code it chooses or no code at all. "Stricter codes have been fought bitterly by politically powerful builders' lobbies," says Clifford Krauss of the *New York Times*, who goes on to note that "the energy requirements in building codes remain weak across half the country, and at least seven states have virtually no rules."[19] In this light, the House of Representatives made history when it passed the American Clean Energy and Security Act (Waxman–Markey bill), which includes a requirement to establish the first mandatory nationwide building efficiency code.[20]

But no bill with national building codes has yet crossed the president's desk, and codes and standards have their limitations in any case. Unseen by most of us, codes and standards go through a long, laborious process with industry code-setting committees or the DOE. Building codes are updated on a three-year cycle; appliance standards can take as many as seven years to get through DOE's process. In addition, enforcing building codes can also be expensive and problematic. Building codes are enforced at the local level in most of the United States. Raising building codes means that both builders and building inspectors must go through a substantial training or retraining period to upgrade construction and practices.[21]

It is an extremely good idea for every state to have a strong building energy code alongside good national appliance efficiency standards. As the House-passed climate bill recognizes, cost-effective codes and standards are necessary elements of a sound climate policy. Nonetheless, the limitations on choice created by codes, the difficulties of enforcement, and political reality all suggest that these policies are better for setting the minimum levels of efficiency that must be offered, not the maximum that can be achieved. There are many cost-effective efficiency opportunities that do not lend themselves to a code-like mandate.

The second good approach is utility energy efficiency programs. Utilities offer a wide variety of programs, often incorporating just about every EE policy tool ever invented—free or low-cost audits, free technical assistance, low-interest loans, and so on. However, the archetypal and most important utility program offers a combination of upfront capital contribution, usually in the form of a rebate, and technical assistance in building a better structure or choosing a more efficient piece of hardware and getting it installed. To cite just two of many examples, Progress Energy Florida offers a program where it will pay $450 toward the addition of a solar hot water heater in your home, reducing your hot water costs by an estimated 85%. Progress offers technical assistance along with the payment, and it works with the Florida Solar Energy Center to certify a group of manufacturers and installers who are eligible to install the systems. In California, the Savings-by-Design program will pay rebates for improving the efficiency of a new commercial building above the levels required by California's already-strict building code.[22]

When they are run well, utility programs can be extremely effective as well as cost-effective. They represent one of the only potentially viable options for tapping our national efficiency potential and thereby reaching our climate change goals. However, utility regulation has not provided clear, consistent, or widespread incentives for this to occur. Creating these incentives is one of the industry's most pressing needs and, consequently, an issue that permeates the discussion in the next several chapters.

The only major alternative to utility EE programs is the provision of financing similar to utility rebates by public entities of one form or another. Low-interest loans and access to revolving funds have both proven effective for government agencies that face chronic budget constraints. Missouri's revolving loan program—one of many good examples—has funded insulation upgrades for public buildings, more efficient street lamps, and automated thermostats for school districts, universities, cities, and counties across the state. The loan payments for this program are often low enough to be covered out of the efficiency savings and are continuously cycled back into more loans. In total, the program reduces the amount of taxpayer money spent on utilities.[23]

Outside the government sector, however, low-interest capital offers themselves have not been successful on a broad scale. By themselves, loan programs do not address the information or transactions-cost barriers at all; in fact, by creating another program that often requires extensive paperwork and credit checks, the hassle factor is only increased. Over the years, many different types of capital access programs have been tried. Some work better than others, and while they are not sufficient, they certainly serve a useful function.[24]

Recently, government EE assistance has taken a giant new step in the form of a new type of program known as municipal energy financing. Through these programs, property owners can borrow at low rates from a designated financing entity for EE improvements, and in many cases, renewable energy installation as well. Loans are secured by property tax liens on houses, which take precedence over mortgage and most other claims on the home (this is what enables attractive interest rates). According to the DOE, there are now five of these programs in the United States, the first of which was the City of Berkeley, California's Financing Initiative for Renewable and Solar Technologies.[25]

Municipal energy financing is perhaps the only EE option besides utility programs that can deploy building energy efficiency rapidly over a wide area, with low-cost access to capital and reduced transactions cost. Combined with other state and local policies, it represents a public-sector alternative to utility EE leadership, as well as a potential complementary approach. However, the option is still quite new, and it is too early to tell how successful the idea will be and how well it can be scaled up to serve the entire nation.

Finally, EE can simply be turned over to the private sector. There are already several dozen for-profit firms known as energy service companies (ESCOs) that specialize in designing and installing EE measures. When working independently ESCOs frequently use a *shared savings* business model. The ESCO raises its own capital and pays for all of the cost of efficient new equipment in someone's structure, such as a new commercial air chiller in an office building. The chiller lowers the power bill of the customer for its entire lifetime, compared to the power bill with the old, inefficient chiller. The ESCO signs a contract that allows it to receive and keep most of the difference between the old and new power bill until it earns back all of the capital it spent on the new chiller, plus a profit. After it has been repaid, most of the bill savings reverts to the customer.

Even more than utility programs, ESCOs and their shared-savings model would seem to be an ideal solution to the main EE barriers. Because they spec, install, and provide the capital for an EE measure, the customer sits back and does almost nothing while an experienced firm does all the work and pays for everything out of its own pocket. The customer sees only a lower power bill with no capital of its own ever needed.

Unfortunately, the economics of the shared savings model severely limits the markets and technologies ESCOs can install and still earn a profit. In commercial buildings, ESCOs find it difficult to get customers to agree to measures and terms that tap the full EE potential of a customer site. Instead, they usually settle for doing the easier measures with rapid paybacks without installing the bigger, longer savings measures. This phenomenon is less of a problem when ESCOs contract with government agencies, who operate under mandates to tap as much of their EE potential as they possibly can. Outside the government sector, however, the for-profit efficiency business can't tap nearly as much efficiency potential as utility or publicly funded programs.

Mandating Utility Energy Efficiency

The hottest new tool for promoting energy efficiency is the *Energy Efficiency Resource Standard* (EERS). The idea is disarmingly simple. Each year, tell every seller of electric power to meet a large fraction of its sales *growth* with EE savings rather than sales. In effect, you simply *require* an electric power seller to do whatever it takes to make sure their customers buy less electricity and meet a savings target.

Nineteen states have already implemented some form of EERS, and at least one federal bill, the Save American Energy Act of 2009 (S.548), proposes to adopt a nationwide EERS. Minnesota, Illinois, and Ohio have enacted laws that require sales reductions that ramp up to 1.5% and 2% a year, respectively, by 2015.[26] S.548 would require electricity and gas retailers to reduce their sales by 15% and 10%, respectively, by 2020. Since we know that EIA forecasts only 9.5% nationwide electric sales growth between now and 2020, S.548 amounts to a mandate that U.S. electric sales decline in the coming decade!

An EERS is a big policy hammer, and it works.[27] However, it is hard to see that this approach is a long-term solution. In my experience, legislated savings targets are seldom set at an ambitious, achievable, and cost-effective level. Instead, they are typically a much lower negotiated compromise (for this reason, S.548's ambitious targets should be seen as an opening bid, not the level that will gain passage of a bill). The challenges of measuring the true potential for cost-effective conservation provide the perfect excuse to keep targets low.

For and Against Utility Involvement

Utilities' current business model is based on selling more power rather than less. Even in states where retail sales are deregulated and utilities only transmit others' power, lower sales mean lower profits. Utilities' understandable ambivalence toward conservation can be overcome either by mandates, as in the EERS, or by changing their business model to make conservation profitable. But before we go about trying to change the core mission of utilities, we ought to ask whether it is the best available means of improving the efficiency of power use.

There's a pretty good case to be made. First, utilities typically provide a place where customers can get extremely up-to-date and unbiased information on the performance and value of EE technologies. Utilities with strong EE programs are some of the best repositories of energy-efficiency expertise in the world and certainly the best in their local areas.

Utilities also have unique advantages raising capital and effectively loaning it to their customers, which is what they do when they give customers a rebate or low-cost loan for efficiency upgrades. It has long been known that energy utilities are one of the most capital-intensive sectors of the economy, and they are adept at navigating between bankers and regulators when they need to raise capital. Because they provide an essential service, and are generally quite creditworthy, they can often raise capital at lower interest rates than most other businesses. Since the value of EE measures depends critically on the cost of capital employed, this is a valuable aspect of utility involvement.

In some ways, utilities are the ideal, and lowest hassle, financing entities for efficiency capital. Utilities serve virtually every single building in their service area. No matter how unusual a building or occupant, they know how much energy it uses, making it easier for them to design measures and estimate savings. They also know the credit history of the customer, as they've been serving them all along with billing and payment arrangements already in place. They have a meter in place to measure savings, and they know it is rare that their customer stops paying their power bill, which is how their efficiency loan is repaid. In short, the captive relationship between the local distribution company and the customer, once seen as a negative in the deregulation debates, is a positive one if the distributor is also the efficiency capital and service provider.[28]

Another advantage of utility pilotage comes from the fact that many customers trust utilities to offer relatively unbiased advice on saving energy compared to most other for-profit companies trying to sell them efficiency measures or loans, including ESCOs. Because utilities are either regulated or publicly owned, market research has shown that customers feel that their advice and the terms of their programs are being monitored by their representatives to make sure they are effective and fair.

Utility-style programs can also provide some help overcoming the hassle factor of installing EE upgrades. The degree to which utilities can offer onsite assistance with upgrades, or even provide them directly, varies greatly by program, utility, and customer segment. At a minimum, however, many utilities prequalify vendors and construction firms. This alone is of enormous value, as the task of finding a well-qualified and reputable installer for many building upgrades is time consuming and often unsuccessful for many owners.

Finally, utility-style programs are the ideal complements to the other EE policy tools. Building codes and appliance standards are essential for creating minimum efficiency levels. However, code and standard designers understand that there are many opportunities to gain savings well above the levels they can mandate. Accessing these opportunities requires the more customized utility-style programs or their equivalent. Moreover, there is a learning cycle that goes on over time. Utility-style programs demonstrate how newer, better technologies can be installed and operated, giving the construction and real estate industry more confidence that these advanced technologies can be mandated in codes without causing great hardship. This cycle of efficiency policies is nicely illustrated in this figure, prepared by Pacific Gas and Electric Company and the California Energy Commission.[29]

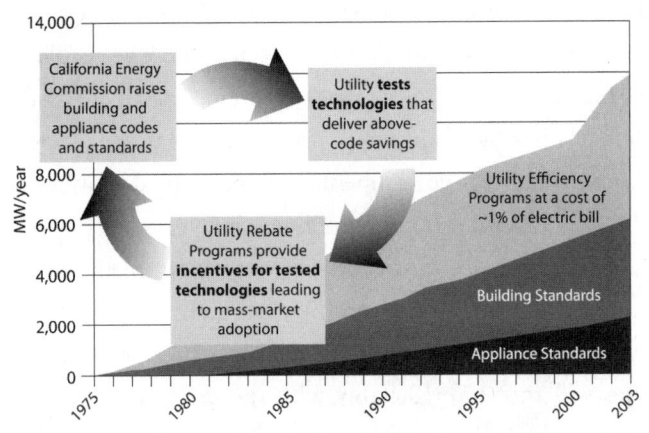

Figure 10-1. How Energy Efficiency Standards and Utility Energy Efficiency Programs Complement Each Other.
Source: Pacific Gas and Electric Company.

The Government Option

There's a nice bit of irony here. Of all of the policy options proven successful at saving electricity, utilities helping customers install EE measures is one of the best, with several unique advantages. When they really try, utilities are pretty good at destroying their own sales! Incidentally, this applies to all types of utilities—investor owned, government owned, and cooperatives. All are highly trusted by their customers for energy assistance and all are capable of mounting very effective programs.

Yet despite these advantages, the industry's fundamental business and regulatory models encourage energy sales, not savings. Programs that are nearly the same as those of utilities can be offered by government agencies, which obviously don't have incentive conflicts, or offered by government agencies but delivered via utilities. In the state of Vermont, for example, a government agency designs all of the energy efficiency programs, and the utilities offer each state-designed program. According to a recent American Council for an Energy-Efficient Economy (ACEEE) survey, of the fourteen states with the best energy efficiency performance, nine rely primarily on utility programs, but five of the leaders put state agencies in charge.[30]

There are pros and cons to getting utilities out of the energy efficiency business. On the plus side, utilities would not face divided incentives, and we would not have to order them around or change regulation. The regulatory and administrative burdens of supervising these programs, which are costly and make the programs less flexible than purely private offerings, would be eliminated. Under the right conditions, governments can raise capital at even lower interest rates than utilities, including tax-exempt financing.

However, there are also downsides to government management of EE financing and deployment. First, government programs can be even more cumbersome to manage than those of utilities. There is also the chance that government policies will shift unpredictably with the political tides. In the case of energy efficiency, the United States has already gone through episodes of this nature. In 1994, when Congress shifted from Democratic to Republican control, one of the first acts of the new Congress was a somewhat successful attempt to stop any activities at the DOE that resembled utility or ESCO deployment programs, such as the program where DOE pays to weatherize low-income homes.[31] The DOE was told to stay out of direct funding of or involvement in any actual installation of efficient technologies (the market would do it). The directive adversely influenced many state efforts as well.

Another downside to government management of the EE function is the amount of capital required. In the current economic climate, government borrowing has become a very important issue. The United States and many other countries around the world have critical needs for public capital to pay for health care and other social services, revive the financial sector, improve public education, provide for national and cyber security, and fund a deteriorating public infrastructure.

The capital needs for EE are substantial, and even with the Obama stimulus package and the relatively small EE efforts of the utility industry to date, the vast majority of EE capital spending has been nonfederal. Until the Obama stimulus package, the total federal budget for EE deployment was well under $1 billion a year; the total of state funding was also of this rough magnitude. The federal government didn't even use its own capital for its own buildings, opting instead for shared savings deals with ESCOs. Meanwhile, utilities are now spending almost $4 billion a year on EE programs and are likely to reach $10 billion in the next few years.[32]

In view of the strong opposition to federal EE in 1994 and the other fiscal needs of federal and state governments, it is difficult to imagine gaining approval for tax increases large enough to fund outlays of this magnitude. Also, if funding like this was achieved based on a precarious political victory, with future funding streams uncertain, the impact on planning and implementing a sound efficiency and climate policy would be devastating. In contrast, regulatory utility EE policies can take a long time to establish, but once they are set they are not usually rapidly reversed.

When it comes to this core element of our climate strategy, we have a critical choice to make. We can order utilities to do it, as we're doing today, against their own business interests. We can take them out of the role entirely, placing the responsibility on our already-stretched state and local governments. Or, we can change regulation and utilities' business mission to give them an incentive to save energy as well as sell it. Any of these approaches might work, but as a nation we aren't doing *any* of the three today, and that's a big mistake.

Although it would be great to fully finance and lead EE actions, I suspect that utility leadership has a slight advantage over giving the job to the public sector directly. As I've explained, utilities cover the entire country, have good capital access, and have the best platform for building a specialized, high-quality delivery system. I also believe that utilities will find much more efficiency to harvest if they can make reasonable profits off it, not just answer to mandates. In the rest of the book we explore what it takes to put utilities in the lead using incentives rather than energy efficiency mandates alone.

CHAPTER ELEVEN

Two and a Half New Business Models

IN ALL INDUSTRIES, but especially in electricity, the issues of economic structure, regulation, and business model are inseparable. Regulation determines which portions of an industry must offer their services as a common carrier or are under an obligation to serve all customers. In these parts, regulation also sets the allowed products, prices, and terms of service. The regulatory and economic structure together establish reasonably well-defined incentives and constraints around which the owners of the industry's assets create business models. The business model represents management's strategy for maximizing the profits allowed by regulation or markets. Figure 11-1 shows the relationship between the three elements of the triad.

In this chapter, we begin to examine the feasibility for creating different triads of structure, regulation, and business models that will facilitate the electric industry's transformation to a smarter future. The industry of the future will work best if all three parts of the triad are mutually reinforcing and consistent with our policy goals.

The discussion is divided into four parts. We first examine the two common present-day structure–regulation–business model triads. I label them by their

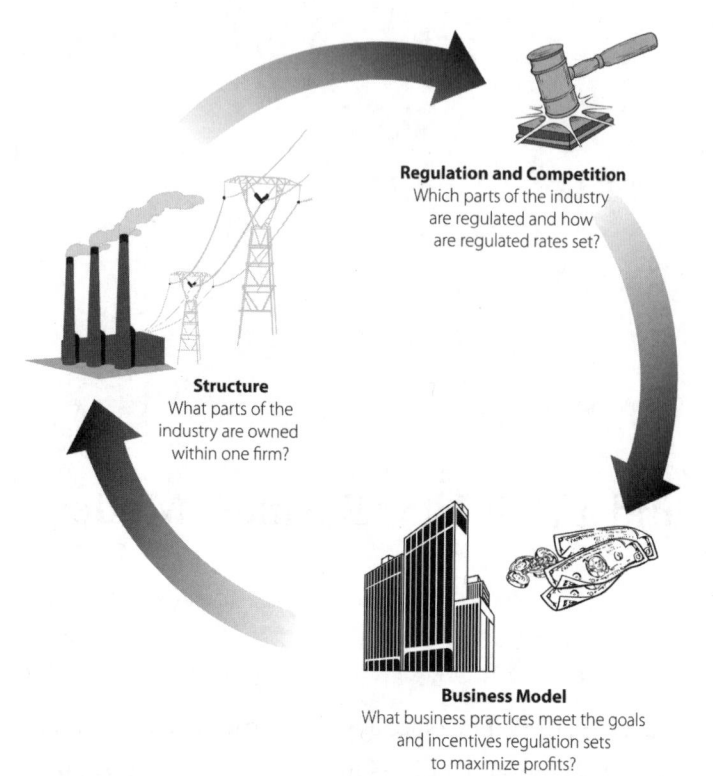

Regulation and Competition
Which parts of the industry
are regulated and how
are regulated rates set?

Structure
What parts of the
industry are owned
within one firm?

Business Model
What business practices meet the goals
and incentives regulation sets
to maximize profits?

Figure 11-1. The Power Industry's Inseparable Triad: Structure, Regulation and Competition, and Business Model.

structure and regulation features, calling them (1) the vertically integrated regulated utility, and (2) the deintegrated structure with retail choice. We then look at two of the most important economic forces that have shaped these triads. The first is vertical integration, that is, the savings that occur when a single utility owns all stages of the electric production and delivery process; the second is the benefits of competition in electric generation, which are similar to the benefits of competition in other private-sector activities. The two triads we observe today are the result of an uneasy compromise between these two forces.

After we understand these two forces we'll consider how the changes we've discussed in technologies, control paradigm, climate rules, and energy efficiency policies affect their balance. I posit that the industry's transformation will lead both of the current triads to evolve. The vertically integrated and regulated utility may keep its structure, but the way in which it is regulated and the corresponding utility business model must be retooled dramatically to promote energy efficiency and a freely accessible smart grid. I call the evolved form of a company operating within this triad an *energy service utility*.

Today's deintegrated utility, functioning in an environment where generation is made and sold competitively, must also evolve. Regulation must make the utility at least neutral toward energy efficiency and reward it for operating the Smart Grid efficiently. I call this evolved utility a *smart integrator*.

The Two Current Triads

The two basic power industry structures that coexist in the United States today are shown in Figure 11-2. The first is the traditional utility that owns generation, transmission, and distribution and sells power at regulated rates. In this structure there is a competitive wholesale market, but it isn't the dominant source of the utility's supply, and wholesalers don't sell directly to retail customers. Although most of the industry is investor owned, this arrangement is replicated by public power and rural electric cooperatives through joint ownership and long-term contracts, as we saw in the case of Jim Jura and Associated Electric.

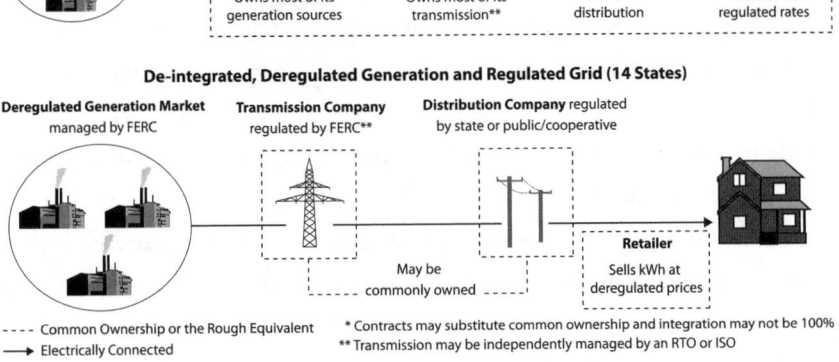

Figure 11-2. Two Industry Structures.
Abbreviations: FERC refers to Federal Energy Regulatory Commission.

The second structure is one where competitive generators are the predominant source of supply and the transmission and distribution networks are common carriers, known in the power industry as open access. These competitors ship their power to customers over large lines regulated by the FERC and distribution systems regulated by the states. Deregulated retail sellers, often owned by the same companies that own nearby generators, purchase power from generators, ship the power downstream, and remarket it to customers at retail.

As mentioned at the beginning of the chapter, these structures are the product of two very significant economic forces. The first is the degree to which the different segments (generation, transmission, distribution, retailing) have the same owner, that is, are vertically integrated. The second is the industry's cost characteristics, which determine whether competition or regulation should yield adequate investment and lower prices. Along with a good measure of political economy, these two forces have shaped the industry we have today, and both will play a pivotal role in the industry's future as well.

Vertical Integration, Old and New

Formally speaking, the power system is a network for several reasons. First, the value of diverse power sources can be maximized, or the costs of providing reliable supply minimized, by interconnecting all sources of power to a single grid with a single control platform. That's why nearly all generators of every type are designed to plug in to the grid, and why there is only one set of wires and system operators in almost every part of the world with electric service.

Network efficiencies also arise from interconnecting customers, each of whose demands vary from moment to moment. When all these changing demands are added together, it is cheaper to provide service to the group than it is to serve each customer alone. This was Insull's massing of consumption in his original business model. As we saw in Chapter 5, the modern-day benefits of networking customers go beyond aggregation of passive demanders to the price- and control-sensitive interaction of customers and system operators.[1]

When there is this kind of interdependence between generators, customers, and other parts of the system, it is often cheaper if many elements of the network are commonly owned. There is an economic theory, attributed to economists Ronald Coase and Oliver E. Williamson, that explains this, but it comes down to the fact that when different parts of a network need to work together in complex ways, it is hard to describe how network parts must work together in a written contract. Because contracts are incomplete and imperfect, the contracting parties don't always work well together when unusual circumstances occur. In contrast, if you own the whole network you manage it as a single company and command any needed adjustments to new developments.[2] Integration is also observed where lack of immediate performance by one small part of the network can be costly to everyone on the network, a feature that fits electric grids to a tee.[3]

Traditionally, nearly all economists agree that vertical integration made the power system work more cheaply.[4] A number of economists who have studied the question prior to the onset of the Smart Grid and related changes have concluded that vertical economies are measurable and quite significant. In one major 1996 study, Professor John Kwoka, Jr., found that vertical integration reduced the costs of the average large utility by 22%.[5] Professor Robert J. Michaels reviewed the literature in 2006 and found strong support for vertical economies ranging from 1 or 2% to Kwoka's double-digit levels.[6]

Over the years I've had occasion to see some of these integration economies in action. I can recall sitting in the control room of a utility one morning looking at the unit commitment computer. This computer indicated which of the power plants was the cheapest one to start up to serve demand over the rest of the day. The utility owned most of its power plants, but a few were owned by others, with their availability determined by the terms of a complicated long-term contract.

When the computer showed that the company's own plants were cheapest to turn on there was no question that these plants would be used. As long as the plant was operable it went to work. When the computer indicated that a contracted unit was cheapest, however, the personnel checked a thick binder that summarized the plant's contract. The contract determined whether, in this immediate instance, the utility had the right to order the plant to turn on for the precise period that would be cheapest. Sometimes it did and sometimes it didn't. Had the utility owned this generator there would be no doubt it would be the one used.

As I've watched power sectors all over the world go through restructuring, one of the things that has struck me is that vertical integration seems to be a very durable feature. In the United States the original onset of retail choice reduced the fraction of assets owned by fully integrated IOUs by about 10%, but since 2003 vertical integration measured by assets has gone halfway back, increasing by about 5%. The reversal appears even more pronounced in Europe, where national deregulation laws forced utilities to sell off their generation businesses. Whereas many of us thought this would mean that distribution systems would not be owned by generators, and that the retailer would be independent of everyone, the generation companies have steadily reintegrated into everything except transmission, the one segment they cannot enter. Today, among foreign electric companies in the world's top 250 largest energy firms, the firms who own generation are either fully integrated or also own distribution systems as well as retailing operations. [7]

The diminished allure of companies operating in just one industry segment—*pure plays* in Wall Street parlance—was explained in a 2002 report by consultants Booz Allen Hamilton:

> Traditionally an industry of vertically integrated companies, the boundaries of which were defined by geography and historical accident, the utility sector has undergone a fundamental and comprehensive restructuring in recent years. Regulatory mandate and market discipline have driven the development of intermediate, wholesale markets and new "pure play" business models in transmission, generation (gas, nuclear), trading, and retail services. The market, recognizing that each of these niches offers distinctive growth and risk characteristics, has generally applauded this trend. Indeed, as recently as a year ago, it appeared that the single most important driver of an energy company's market valuation was the strategic position it elected to occupy on the industry's value chain. But that was a year ago. . . .
>
> Since then, the ground has shifted, and many of the assumptions underlying energy companies' strong valuations have been shaken. The industry lost $90 billion in market value in 2001, and continues to hemorrhage in 2002. Falling energy

prices, looming overcapacity, and a crisis of confidence have cast a pall over merchant [deregulated] energy companies and generators. The California energy crunch and the Enron debacle have stymied the progress of deregulation. And investors have stepped back from their lofty endorsements of "intellectual capital," "intangible assets," and energy "pure plays." In fact, investors appear reluctant to bet on any particular strategic posture these days. As a result, average valuations assigned to different "plays" on the utility value chain have converged, while, at the same time valuations *within* particular strategic niches have spread.[8]

While the cost savings from vertical integration can be substantial, neither should they be blown out of proportion. There are many, many instances in the industry in which contracts do an excellent job of substituting for common ownership and many others in which they do better. As in a host of other industries, cheap and ubiquitous information technology (IT) has made it possible to turn a group of different companies and their assets into a single production operation that works like a single company, even for industries in which continuous service in real time is essential.[9]

More importantly, though, all the experience and evidence to date apply to yesterday's electric power industry. The sixty-four-thousand-dollar question is whether the smart utility of the future will enable cheaper total service to customers if it is vertically integrated. Would it instead be cheaper to customers to avoid cross-ownership between generation and grid and operate only in one or two of the segments?

The issue is far from settled. It is possible that the economic attributes of a more distributed and intelligent smart power network make it easier for the transmission and distribution operators to balance the system and keep costs low without needing to own power plants. By sending price signals to generators large and small, they can get all the generation and all the control they need when and how they need it. On the other hand, it may be that the vertical integration economies found in the old industry structure grow larger due to "subtle and complex" factors we can't yet foresee.[10]

One result is certain. As we saw in Chapter 5, the policies regulators or public owners adopt regarding pricing and other aspects of the Smart Grid will have an enormous impact on the economics of all other parts of the system. It is thus unavoidable that these regulatory policies will do much to determine whether integration is cheaper and/or more administratively feasible than separate ownership. Legislators or regulators also simply may *impose* a structural outcome. This is exactly what state laws do in the fourteen states that have mandated retail choice, where state laws forbid the owners of distribution systems from also owning in-state power generators.[11]

In short, the winner of the tug-of-war between vertical economic efficiency and regulation will determine whether integration survives the Smart Grid. To assess the odds, we turn to integration's countervailing force, competition.

The Benefits of Competition

Nearly everyone familiar with the industry remembers that utility regulation arose out of a belief that power companies were natural monopolies. Natural monopolies are characterized by economies of scale, that is, average production costs go down as output goes up. The more precise, modern term for these cost characteristics is *subadditive costs*—those where the cost of producing 200 units of a product is less than twice the cost of producing 100 units.

In Figure 11-2 there are four segments of the industry: generation, transmission, distribution, and retailing. When regulation was applied to the industry, no distinction was made between these segments; all segments were regulated. A century later we know that natural monopoly attributes apply differently to each industry segment. Transmission and distribution are still considered natural monopolies largely because it would be extremely intrusive, and almost certainly inefficient, to have two sets of grid wires sitting side-by-side. It would be like paying to dig two independent sets of ponds and canals, both running to all the same locations.[12] And two might not be enough for strong competition.

The cost characteristics of the generation segment are much more complicated. It is often thought that a natural monopoly in generation means that a single large power plant is cheaper than several smaller plants. This is still true up to a point, but plant-level scale economies have pretty much topped out. It is no longer true, for example, that a 2,000 MW traditional power plant is cheaper than a 1,000 MW version. In fact the average size of a traditional power plant added by utilities peaked at 493 MW in 2003 and has since fallen steadily to 171 MW in 2008. While this size trend is undoubtedly the product of other factors in addition to scale effects, it does suggest that the cost dominance of large plant economies has come to an end.[13]

If it is equally cheap to own fifty medium-sized plants versus five ultralarge ones, there is still the question of whether it is cheaper to have fifty companies, each owning one plant, or five or even fewer generating companies owning the fifty plants. In economic jargon, this is called *firm-level economies of scale* (as opposed to *plant-level scale effects*). This is roughly equivalent to asking whether the industry is an inherently competitive group of firms, a natural oligopoly (several sellers), or a firm-level natural monopoly.

This leads directly to the two options in Figure 11-2. If a single company owning all generators (whatever their sizes and types) is most efficient, it is a true natural monopoly. In this case, you would probably want to establish and regulate a generation monopolist. Once you do, you might as well let them integrate into transmission and distribution, which are also natural monopolies, and then regulate the whole thing, because this way they can also gain the cost savings from vertical integration. This reasoning is the motivation for the integrated regulated utility triad. If regulation itself does not introduce too many inefficiencies, this ought to be cheapest, while protecting customers against monopoly abuse via regulation of rates.

The other possibility is that it is cheaper to let multiple generating companies, each owning enough generators to achieve a low-cost operating scale, compete with each other to sell downstream. This approach is cheaper if competition forces these companies to find more efficient ways to build and operate plants than they would if they were regulated. Unleashing generation competition in this fashion would mean giving up some or all of the cost savings from vertical integration, because many of the cost-savings practices from vertical integration interfere with, or are removed by, generator competition.[14] But if the cost savings gained from competition outweigh the savings lost from integration this is a good tradeoff to make.

The latter proposition reflects the thinking that led to deregulation in the 1990s. Multiple generation companies could own enough generators to achieve firm-level economies of scale, but there would still be a large enough number of companies to compete to sell power. For example, there might be only three or four generation companies in any geographic area, but in many cases this is enough to keep prices at roughly competitive levels (in other cases, it isn't, which is why competitive electric markets need to be watched carefully). Companies that owned transmission and/or distribution would no longer be allowed to be vertically integrated (i.e., own generators), or if they did they would not be allowed to give them preferential treatment over competitive generators they did not own.

Will Deregulation Blossom under the Smart Grid?

As we learned in Chapter 2, the California power crisis and states' experience with provider of last resort (POLR) rate increases has reversed the adoption of retail choice. While a few states with retail deregulation are still debating whether to turn back the clock, most of the current debate centers on whether generation competition has worked at the wholesale level, and how wholesale markets should be designed to make them work better.[15]

The transition to a smarter and more distributed system adds a new chapter to this ongoing debate. Proponents maintain that the coming more decentralized structure is inherently better suited to generator competition. Indeed, many Smart Grid proponents suggest that the benefits of retail choice and the Smart Grid are integrally related.

There are two strains to this line of reasoning. The first looks broadly at our experience with wholesale competition and the basic incentives of integrated utilities. For example, many of the new generating technologies needed to meet climate targets require innovation and risk taking by their owners. Proponents note that these cost and performance risks are better handled by a competitive generation sector because unregulated firms are run less conservatively, funded by capital willing to take risks, and more experienced with innovation than regulated and publicly owned utilities.[16] They also argue that regulated utilities are resistant to unconventional generators because any outside source of power tends to reduce their market share and ultimately slows their asset growth.

These proponents also note that deregulated wholesale markets in the United States already have prices that vary not only by hour but also by location on the power grid. While these *locational marginal prices* (LMPs) are unpopular with some stakeholders, they increase revenues for competitive generators that produce during high-price hours or can sell at high-priced locations on the grid. These price signals and the receptivity of customers and grid operators to greener and more innovative sources lower the barrier to low-carbon supplies in competitive markets. They note, for example, that three times as much wind energy is made in organized wholesale markets as in areas without them.[17]

There is room for disagreement with this overall view. As an example, several of the states with extremely forward-looking climate policies and strong commitments to renewable energy—California, Colorado, and Nevada, to name three—do not have retail choice, and only one of the three operates in a centralized wholesale market. However, there is one important and undisputed reason why generation competition will work better when dynamic pricing becomes more widespread in its use: unlike today, demand will change with price.

One of the biggest problems with unregulated competitive electric markets has been the fact that supply and demand cannot adjust the way markets are supposed to "equilibrate" in the theory of competitive markets. In textbook markets, when prices rise supply expands because new suppliers come into the market and existing suppliers boost their output. Of equal importance, when prices rise customers cut back on their demand. The combination of higher supply and lower demand brings prices back down.

In the absence of dynamic pricing it is not so surprising that these markets haven't worked as well as they could—half of the participants in the market *couldn't* respond the way they should.[18]

The greater participation of customers in electricity markets is the centerpiece of the second strain of thought on retail choice and the Smart Grid. "Retail electricity markets must integrate consumers as active partners in balancing the electricity supply/demand equation," write Robert Galvin and Kurt Yeager, former executives of Motorola and EPRI who founded a Smart Grid incubator. In their view, markets with retail choice will unleash innovation and a drive for quality that regulated utilities have long resisted.[19] Many members of Smart Grid trade groups agree, though they often speak in terms like "greater customer control" rather than referring to deregulation directly. One leading Smart Grid economist, L. Lynn Kiesling, writes that

> Decentralized coordination is good, and is preferable to centralized control because it harnesses the dispersed knowledge of many market participants, it honors differences in individual preferences, and it enables discovery of individual preference and cost differences, through differences in consumer willingness to pay and producer willingness to accept. Decentralized coordination leads to more robust and resilient economic efficiency in the face of change and has more adaptive capacity over time than a system that relies on centralized control.[20]

Inherent in this strain of thought is the idea that the hardware of the Smart Grid will enable many new valuable energy-related applications that customers can load into their computers or home energy management networks, much like applications can be downloaded into our cell phones.[21] The fear is that if regulated distributors have too much control, innovation will be squelched.

However, it is not automatic that a single supplier of generation at dynamic prices rather than many suppliers would negate the incentives for new application development. Florida Power and Light (FPL), a utility that does not operate in a retail choice state, has announced that it intends to launch a smart grid that will feature an open applications development platform. Software developers can offer FPL's customers new ways to track and control their energy, uploading their programs onto FPL's customer software platform, much like iPhone application developers upload their applications onto Apple's iPhone system. At least in Florida, the intelligence will come from the marketplace, but the electrons will all come from FPL.

While there is little doubt that competition will perform better under dynamic pricing, it is far from clear that the industry's technical transformation and further deregulation are as essentially linked as these proponents argue. It is entirely possible that the profound changes under way will compel further deregulation simply because the industry cannot finance or operate the new infrastructure solely as a collection of regulated generation owners. However, it is also possible to enable huge amounts of customer optionality and control even if the local utility is the aggregator and balancer of all power sources. It is too early to tell, but the early signs are that it is just as possible that retail deregulation will not expand in the near term, at least until the Smart Grid is better established.[22]

Structure and Regulation Futures

The possible outcomes for structure and regulation as the vertical integration and competition unfold are shown in the two-by-two matrix in Figure 11-3. The columns on the chart distinguish between two futures in which the network effects that have led to vertical integration continue to be strong (left column, Quadrants I and III) or weaken (right column, Quadrants II and IV). In the left-hand column, vertical efficiencies increase in importance; in the right-hand

| | | Smart Grid Network Effects . . . | |
		... Favor Vertical Integration	... Favor separate ownership of generation and grid
Impact of Competition	Benefits of generation competition increase under smart grid	**Quadrant I.** Conflicting economic forces could lead to either more or less integration and deregulation	**Quadrant II.** Generation and retail sales are deregulated as in current retail choice states; transmission and distribution regulated and operated as smart grid platform
	Benefits of generation competition same or less	**Quadrant III.** Integrated, regulated utilities continue to sell generation and operate the smart grid	**Quadrant IV.** Conflicting economic forces could lead to either more or less integration and deregulation

Figure 11-3. Future Industry Structure and Regulation Outcomes Are the Product of Two Forces: Network Effects and Competition.

column they decline. The two rows stand for the benefits of downstream competition for generation and related services. In the top row, the benefits of generation competition increase as part of the transition, as the proponents of deregulation believe will occur. In the bottom row, generation competition stays about where it is today, divided between states with and without retail choice and shaped by FERC's rules in the wholesale market. Overall, each quadrant of the matrix represents one possible structural/regulatory future, each characterized by the differences in the value of vertical integration and generation competition.

Of these four futures, the two important ones are in the northeast and southwest quadrants, numbers II and III. Quadrant II is one in which vertical integration benefits are not strong, and competition among more decentralized generators works well. This scenario corresponds pretty closely to the future vision of the industry espoused by deregulation proponents, Smart Grid enthusiasts, and some utilities. In this future, utilities stay out of the business of generating power and limit themselves to running a smart transmission and/or distribution system that integrates, sets prices for, and balances all types of generation, storage, and demand response. Utilities become what I've called Smart Integrators.

In the other important scenario, Quadrant III, the value of vertical integration stays strong, and generation deregulation is not more attractive than it is today. This future is one that looks a lot like today's traditional vertically integrated and regulated utilities, except that they now use Smart Grid technologies, dynamic pricing, and decentralized sources. These companies become Energy Service Utilities.

In the remaining two quadrants there are conflicting economic forces. In Quadrant I integration benefits are strong, but so is the value of generation competition. It isn't clear which provides more cost savings, nor which one legislator or regulators will favor as the Smart Grid is implemented and the nature of the two forces becomes clearer. The same is true in Quadrant IV, where vertical integration isn't highly valued, but neither is competitive generation.

While either of the two conflicting quadrants may end up being the most accurate description of reality, I don't believe they represent long-term industry

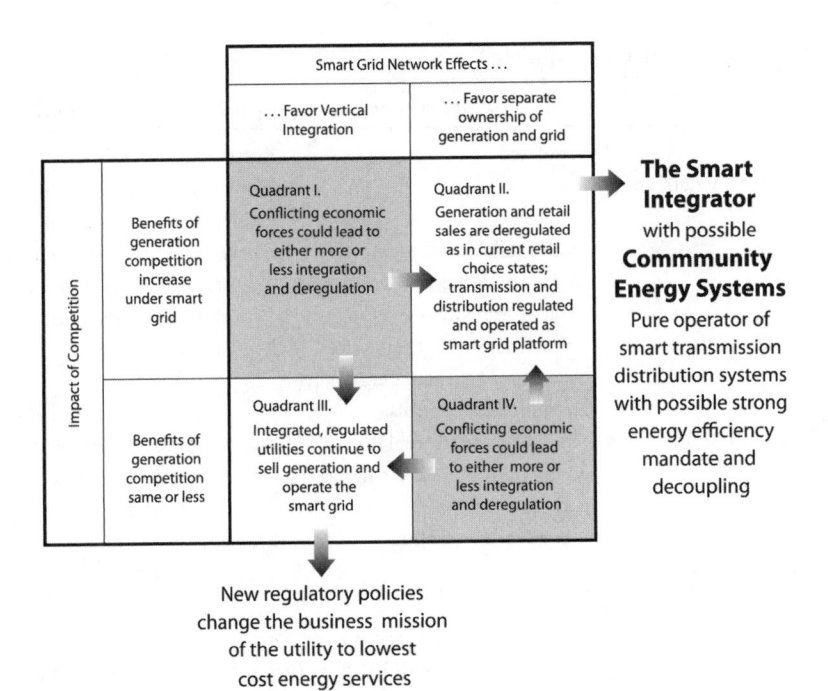

Figure 11-4. The Two Likely Future Industry Business Models.

outcomes. In relatively short order, I think that regulators, policy entrepreneurs, or stakeholders within the industry will resolve the conflict by moving the path of the industry into Quadrants II or III. Figure 11-4 shows this prediction by adding arrows out of the conflicted quadrants into the more likely futures.

The result of all this are the two main business models–structure–regulation triads summarized in Table 11-1. The Smart Integrator business model is a utility that operates a regulated smart grid offering independent power and other services at market prices. Its energy efficiency role is up for grabs, but its disincentives can be removed by decoupling. The second business model, the Energy Service Utility, is vertically integrated, regulated, and must have strong EE incentives built in to its regulatory structure to offset its regulated profit motive.

Standing back to look at this result, we see that this whole discussion has led to the unremarkable conclusion that the two likely future triads are quite similar to the two that we already have in the United States today, with the Smart Grid

Table 11-1. Two Business Models

	Smart Integrator	*Energy Service Utility*
Retail Prices	Deregulated	Regulated
Generator Ownership	None	May own some or most of its supply
Power Delivery Role	Operates reliable balancing, delivery, and integration network to all power sources	Operates reliable balancing, delivery, and integration network to all power sources
Price-setting Role	Operates market that sets prices or within a market operated by an independent entity	Regulators set prices
Information Role	Operates information platform giving customers price signals and controls	Operates information platform giving customers price signals and controls
Energy Efficiency Disincentives	Removed by decoupling	Efficiency is made a core mission and profit center

and its accoutrements added on. This outcome is not surprising. The power industry can be justly criticized for resisting some types of change, but it is also an industry that must do all its structural and regulatory alterations while the plane is in the air. In the wake of the California crisis and subsequent events, there is limited appetite among utility policymakers to pursue bold, untested changes. We should make it our goal to accelerate the pace of change, but I do not think we can change the incremental nature of the path.

Two and a Half Business Models

We last come to the half-business model referred to in the title of this chapter. The half refers to a variation on the Smart Integrator scenario in which distributed generators are owned by their communities. Otherwise the rest of the scenario is the same.

Distributed generation (DG) scenarios are often presented as if most small generators will be owned by individual homeowners and businesses, but this is far from certain. While scale economies no longer favor ultralarge power plants, it is still much cheaper to install medium-size distributed generators too large for one customer rather than very small ones. This is demonstrated in Table

Table 11-2. Community-Scale versus Individual-Sized Distributed Generators

	Typical Size		Approximate Current Cost* (¢/kWh)	
	Small Unit	Medium Plant	Small Unit	Medium Plant
Combined Heat and Power (Cogeneration)**	20 kW (Microturbine)	50 kw–20 MW (Industrial cogeneration)	8¢	4¢
Wind Power	5 kw	5–20 MW		6¢
Photovoltaic Solar	4.5 kw	10–20 MW	34¢	17¢
Fuel Cells**	10 kw	2 MW	19¢	17¢

Notes: *Excluding backup and integration for variable sources
**Natural gas prices are assumed to be $6/MMbtu
Sources: see Appendix B.

11-2, which shows that the price of power from DG typically doubles or more at a household-size installation. Recognizing the superior economics of midscale resources, the National Renewable Energy Laboratory notes the following:

> Integrating the Renewable Energy Community as a whole system can accrue significant benefits. Cost advantages from the systems approach—linking homes with vehicles and addressing energy issues on a community level rather than on individual households—can be gained compared to the costs of each individual part.[23]

The size cost disadvantage can be reduced if hundreds to thousands of individual customers all install DG using standard practices, but the financing and management of this network of tiny resources requires an administrative and financial infrastructure of its own.

Private, deregulated companies will certainly seek this market out. In a state where the local distributing utility cannot own generators, the likely nonutility owners of midscale DG are the independent power companies already in the business, such as AES Solar and SunEdison. As an alternative, however, the prospect of economical midscale power has also renewed interest in a new form of public power called community energy systems (CESs). A CES is a publicly owned entity that owns locally located medium-scale electricity sources and possibly energy storage or its own distribution wires. It sells the power it makes, along with storage and other related services, to its customer-members.

Where the distribution system is already owned by the community, as in about 2,000 cities and towns across America, the local public power company can become the CES. However, this returns the structure back to vertical integration, a model we'll explore in the next chapter. The more unusual possibility is that a desire for community control and resistance to deregulated power sellers will create an upsurge in nonprofit CESs where the distribution system is investor-owned and barred from owning DG itself. The Galvin Electricity Initiative reports increased activity in this area from communities and other large institutions such as universities. The public power community has also taken notice and is promoting its own involvement in small-scale energy, albeit often integrated with wires ownership.

In the next two chapters we take a closer look at the Smart Aggregator and the Energy Service Utility. The good news is that all two and a half main business models look capable of providing reliable service while the industry goes through its coming technological upheaval. The bad news is that neither of the models is proven, and experimentation with new regulatory and business models is always difficult while the plane is in the air.

CHAPTER TWELVE

The Smart Integrator

THE SMART INTEGRATOR (SI) is a utility that operates the power grid and its information and control systems but does not actually own or sell the power delivered by the grid. Its mission will be to deliver electricity with superb reliability from a wide variety of sources, from upstream plants to in-home solar cells, all at prices set by regulator-approved market mechanisms. In addition to the physical maintenance and repair of the local wires (and often upstream as well), it must keep all generation plugged in to its system in balance with demand and its customers fully empowered to shift their use in response to price signals.

The customers enrolled in Northeast Utilities' (NU's) Smart Grid pilot program are getting an early glimpse of a Smart Integrator utility at work. NU's utilities serve in New England, where every state except Vermont has retail choice. NU's Smart Grid pilot customers pay dynamic prices for their retail power, though the power is not purchased from NU itself—customers choose from among a number of deregulated retail suppliers. What they do get from NU are the price signals that are sent to their programmable thermostats, prices sent to their central air conditioners, a monitor that displays the costs of their household power purchases in real time, a Web site with additional services, and an "energy orb" that changes colors as the price of power changes throughout the day. The orb glows green when hourly real-time prices are low, yellow when they are medium, and red when prices are high. In the future, NU customers will be plugging their hybrid cars and solar panels into the same smart grid, selling power back to third parties at prices set by NU's market software.

The physical and information technology (IT) architecture of the Smart Integrator will be designed to fulfill such functions. There will be no change in the utility's obligation to ensure that the local grid can deliver all power that is demanded, and it responds to local blackouts as quickly as possible. It will continue to own the local lines, operate fleets of repair trucks, and respond to power delivery complaints.

This part of the new mission is quite traditional. Every distribution company (often referred to as a distco) already operates a control center much like the larger ones run by the high-voltage system operators (large utilities, independent system operators, or regional transmission organizations). In it, they monitor every one of their distribution circuits and watch for lines going down or other local outages. They communicate frequently with the large-scale operators upstream, especially when problems occur. As we saw in Chapter 4, Smart Grid systems offer the prospect of substantial improvements in these balancing and reliability maintenance operations.

In this scenario, however, the Smart Integrator must attract, ensure, and provide transportation for many generators, small and large. To do this, it will need to provide the most open architecture possible for power sources of all types. Each part of the system where generators might locate must be physically and electrically designed to allow for flow into and out of the grid, a task in itself for systems that are now designed solely for one-way flows. Since the utility will not be able to direct where generators locate and how large they will be, planning the system's capacity expansion to accommodate two-way flows will be a large and continuing job, with enormous uncertainties attached.

The Smart Integrator will also need a second, highly secure but maximally open platform for information, price, and control signals. Smart appliance manufacturers, power management software vendors, community energy systems, and others will all want to send and receive the information they need to decide whether to sell or buy on an hourly basis (actually, the most advanced current systems reset price every five minutes). Along with all this information provision there will be massive new accounting, billing, and settlement systems to accommodate the much more complicated pricing and payment options offered. This will be greatly facilitated by standards and interoperability protocols, but as we saw in Chapter 5, this daunting objective is likely to take time.

The last Smart Integrator task will be the most difficult of all. This utility will have to administer a market that determines the hourly (spot) prices its customers use for trading. In addition to setting spot prices, policymakers will probably insist that the utility offer some sort of provider of last resort (POLR) service for retail customers who do not make or buy their own power from deregulated sellers. As you'll recall from Chapter 2, POLR service was one of the main downfalls of retail choice; it will hopefully not repeat the performance in the coming Smart Grid era.

Finally, unless Congress changes the Federal Power Act, the rules and mechanics for setting the local price, as well as the price for POLR service, will be decided by state regulators for investor owned utilities (IOUs) and managements for public power and cooperatives. Establishing these price-setting mechanisms is a challenging task.

Setting Local Prices

The idea of using many sellers and buyers of power to establish electricity competitive spot prices was settled quite some time ago, and it is used in many "deregulated" electricity markets today (the quotes around deregulated to remind you that there is still a huge amount of regulation in these markets, it is just that regulators don't set prices in advance).[1] The largest electricity market in the world, the PJM market in the U.S. mid-Atlantic region, sets prices for over eight thousand separate locations on the electric grid every five minutes, and it monitors every single price and location to make sure the price determined by its software is the result of adequate competition.[2]

There is every expectation that the basic software systems that take supply and demand offers, and from them determine a market-clearing price, can be used for local smart power markets. In fact, this is essentially what the Pacific

Northwest Labs researchers did in the Olympic Peninsula experiment. However, establishing the IT platforms that set prices in the wholesale markets while keeping the system functioning reliably has cost the RTOs and ISOs that administer them hundreds of millions of dollars. Ten years after the first of these markets was created, the systems are still exceedingly complex and hard to fix. Pushing this level of complexity into every distribution system will be a considerable challenge.

One aspect of the local price-setting process sure to gain regulators' attention is market power. In any market, there need to be several buyers and several sellers in order to make competition robust and the resulting prices fair. Depending on the geography of the local grid, there may be places where there are very few sellers of local power. To prevent these sellers from having too much market power, local regulators will need to create backstop power supplies with controlled prices or impose price controls on local sellers. Both approaches have proven necessary in much larger wholesale markets, though especially the latter. In the most advanced bid-based spot markets, FERC rules now require that the price-setting software automatically check to make sure that there are enough buy and sell bids to make every node in the market reasonably competitive. If there aren't, the price-setting computer automatically limits the prices that can be bid. There must also be a full-time market monitor who examines competitive issues and complaints. Analogous rules and procedures will be needed for local markets.

Spot prices for energy are only part of the value of electricity supplied or saved, and they are the easy part. As we saw in Chapter 5, the majority of the value of demand response (DR) or distributed generation (DG) comes from its ability to defer or displace expansion of the upstream grid and generators, or its capacity value, along with other even more abstract benefits. As we also saw, the value of these benefits often cannot be set easily by markets.[3] In this case, regulators must engage in lengthy proceedings to set methods of measuring the value, and then utilities must administer them under the critical eye of regulators and stakeholders.

Finally, in Chapter 5 we noted briefly that customers will supply some of their own electricity service but will want to buy whatever they cannot self-produce from the Smart Integrator. This service, traditionally called utility backup power, is difficult to price without a number of case-specific calculations. For example, one customer may need backup power only once or twice a month for a solid day or two, while another may need it on an ongoing but highly unpredictable basis. Proceedings for setting fair capacity prices for customers like this are especially hard-fought because customers who rely on the Smart Integrator for all of their supply do not want to subsidize customers who sporadically rely on the power grid only when they need it, and vice-versa.

Core Competencies for the Mission

The number of regulated responsibilities the Smart Integrator will hold suggests that, whatever its mission statement, it will be a creature greatly beholden to its state regulatory commission or if public/cooperative, its governors. It will be subject to many of the political and stakeholder-driven forces that now buffet state regulatory proceedings. It will need to keep its skills at regulatory negotiations extremely well honed, as they will be in constant use.[4]

The power delivery–balancing–reliability maintenance roles of the Smart Integrator will evolve naturally from similar roles distribution utilities play today with less system intelligence. Although the Smart Grid will make this work far more sophisticated than current practices, the objectives will be the same, and upgrading both the information and human resources seems to be a manageable challenge.

Smart Integrators will need to build especially strong core competencies at the evaluation and operation of IT systems and software. In this area, utilities seem to have a long way to go. One software expert, Ali Vojdani, offers this sobering assessment:

A comparison between the business planning environment in the utility and telecom or airline sector shows the contrast between flexible versus static design. Telcos and airlines can change their entire pricing structure in a matter of hours in response to changing business conditions or an advertising campaign by a competitor. For example, when a major airline announces that it will introduce a fuel

surcharge or collect fees for checked luggage or on-board food service, the entire industry typically matches in a matter of hours. . . . Utilities, by contrast, take months,
if not longer, to introduce a new tariff or adjust an existing one, and this greatly
hampers their ability to respond to consumer demand and changing requirements.[5]

Smart Integrators will also need to get extremely good at evaluating the economics of new system and platform investments. When any new investment or
function for the system is proposed, regulators will want to know why it wasn't
included the last time a similar investment was made. They will undoubtedly
ask many tough questions, including whether the current system should be revised or discarded. They will also want to know what may be next in the investment pipeline.

The scale of the IT systems transformation and information management
tasks should not be underestimated. Two Microsoft utility consultants note that
the software utilities use for balancing generation sources will expand from a
few hundred generating sources at most to several million sources in a highly
decentralized grid, necessitating what they call "Internet-scale data acquisition."
The number of control points in customers' homes and offices eventually linked
to system operators will be vastly larger, creating a huge information management task. The Electric Power Research Institute estimates that the amount of
annual data utilities must process will increase tenfold, from 100 terabytes of
data today to 1,000 terabytes of data (1 petabyte) when the Smart Grid is fully
operational. The Microsoft consultants remind us that "this data is meaningful
for decision support only if there is a way to query and analyze it."[6]

The human capital needed to establish and maintain the new core competencies is perhaps the single largest management challenge to this business
model. A typical regulated utility public power agency or cooperative is pre-
dominantly staffed today by electrical engineers. Many of them have a good understanding of IT and/or economics. However, the Smart Integrator operating
in a significantly deregulated local setting is going to need dramatically greater
expertise in IT and regulatory economics. It is hard to imagine a power company with as many economists and IT experts as power engineers, but that is
probably where Smart Integrators are headed.

To Sell or to Save?

Throughout the United States the predominant method of setting the rates for regulated utility services is *cost-of-service (rate-of-return)* regulation.[7] A very similar approach is used for public power and cooperatives, although their profit incentives obviously differ. The Smart Integrator's rates for distributing power and all other services will be regulated, initially in this same manner.

The core principle of cost-of-service regulation is that revenues earned should equal prudently incurred actual costs for the service provided plus a fair return on prudently invested capital. Arithmetically,

Revenues = Costs + Return on Invested Capital.

Revenues for any product equal its average price times the quantity sold. To determine the average rate necessary to provide adequate revenues (the "revenue requirement"), regulators divide the required revenues by forecasted power sales. In other words, they work backwards, dividing the revenue they want to achieve by the quantity they expect to sell to get the average price the utility needs. They later allocate the revenues into customer charges and inclining or declining quantity schedules, and adjust prices up and down to various customer classes, but through it all they preserve the necessary average rate that gives the right level of revenues at expected sales.

The sales number used in this formula is sometimes last year's sales and sometimes a forecast of next year's sales, but either way it is a single number plugged in to the formula to set the rate. Once the average rate is set, it applies until the process is redone, typically two to five years. So, for example, if sales are predicted to be 1,000 kWh and required revenues are $100, the average rate is set at 10 cents/kWh ($100/1,000 kWh) until it is changed.

The simple act of setting a fixed, per-kWh rate and then leaving it there until the next rate proceeding encourages larger sales by utilities and equivalently discourages their energy efficiency (EE) efforts. As long as rates are set per kilowatt-hour, the more kilowatt-hours you sell once the rate is set the more revenue you earn. Since every kilowatt-hour's revenue includes a bit of profit, the more you sell, the more profits you earn. The sales incentive is largest for utilities, who give their profits back to investors, but it also applies to public power and cooperatives, who rebate their profits back to their public- and customer-owners, respectively.

Here is a numerical example. Suppose a utility's required revenues are $1,000, $900 for costs determined to be prudent and $100 for reasonable profits. Regulators expect a utility to sell 10,000 kWh, so they determine an average rate of 10 cents/kWh ($1,000/10,000 kWh). If the utility sells exactly 10,000 kWh it will earn its exact profits. If it gets lucky or is able to promote its own sales and sell 11,000 kWh, it will earn $1,100 in revenues. When it sells 11,000 kWh, its actual costs are higher than the $900 regulators used to set the rate, as the utility had to buy a little more power plant fuel than regulators guessed they would when they estimated $900 in costs and 10,000 kWh sales. A good guess is that costs only went up $50 when its sales increased from 10,000 to 11,000 kWh.

At the higher sales level, its revenues are $1,100, its costs are $950, and its profits have gone up from the reasonably determined $100 to $150, an increase of 50%. This increase did not require any approvals, new investments, or new rates set by regulators—it is inherent in the process of setting a fixed per-kWh rate before you know the actual level of sales a utility will have, or its costs, during the time the rate will be in effect.

This sales incentive is largest and most important if a utility's business model is to own and sell power from generating plants, large or small. It is obviously also true for deregulated generation companies, since their unregulated profit also increases with greater sales. For vertically integrated companies their generation assets are usually their largest assets by value, so they contribute the most to profits. But the sales incentive also exists for the Smart Integrator, since it too earns more revenue and therefore more profit for every additional kilowatt-hour it *delivers*.

Decoupling and Energy Sales Incentives

If you've read much about the utility industry lately you've undoubtedly heard of revenue *decoupling*. Decoupling, which breaks the link between sales and revenue, has been promoted heavily by EE and environmental advocates because it acts to make a utility's profits neutral with respect to energy sales. It is a modification state regulators make to traditional rate setting that works well for Smart Integrators.

To see how it works, remember that the imaginary utility in the last section was expected to sell 10,000 kWh in a year. The utility, whose costs plus profits were $1,000, was allowed to charge 10cents/kWh to earn $1,000 (10,000 kWh × 10cents/kWh = $1,000). The utility had the incentive to boost its sales to 1,100 kWh, which raised its sales revenues to $1,100 and its profits by 50%.

Decoupling simply says to the utility, "If your energy efficiency efforts in the next one or two years change your sales below 10,000 kWh we will still give you $100 in profits in your revenues." So if the utility's EE efforts cause it to sell 9,000 kWh, and its profits go down by the normal arithmetic, then the utility gets to automatically add a special surcharge to the 9,000 kWh sold so that it earns just enough extra revenue to bring its profits back to $100.[8] For this particular year, its profits are $100 no matter how well it does reducing its own sales through efficiency.

This is an improvement over doing nothing, but it is something of a short-term fix. While the utility hasn't lost its $100 in expected profits, its efficiency efforts have now lowered its sales to 9,000 kWh. Investors will see the company as one whose sales are declining, not growing. This is rarely attractive on Wall Street regardless of whether regulators have guaranteed this year's profits. Slower growth means the utility won't be needing to build more plants or lines soon, which means that its invested capital and its capital-driven profits won't be going up either. All other things equal, investors discount the value of utilities who face strong EE mandates even if they have decoupling, because the long-term trajectory of utility capital investments is lower than it would otherwise be.

Smart Integrators, whose only business is operating delivery and management platforms, won't care as much if fewer power plants are going to be built in their area. Lower amounts of power sales do mean lower amounts of power transported and delivered over their lines, but this doesn't harm their profit growth prospects the way it hurts generation firms. Much of the Smart Integrator's asset growth will be needed to provide reliability, replace aging equipment, or make Smart Grid investments that improve operations, none of which are especially sensitive to sales levels. In addition, if the sales of kilowatt-hours delivered go down a lot, and with them the revenues required to earn fair profits, the company can ask regulators to increase revenues by allowing it to charge for other services, such as fees for interconnecting with its network. Regulators will be setting prices for all of the value-added services the Smart Grid enables in any case, leaving many new product revenue streams the Smart Integrator can tap to achieve profitability—all set by the regulators who try to provide the overall right level of profit for all prudent investment.

Wall Street and the Smart Integrator

In the last chapter we saw that the aftermath of the California crisis removed much of investors' enthusiasm for companies in only one power industry segment. As the industry's transformation continues and Wall Street takes a look at the numbers, the Street may recapture some of its enthusiasm for transmission and distribution system pure plays, if not for each of the four segments alone.

The investment community has long viewed the transmission and especially the distribution segments of the utility business as the sleepy, slow-growth parts of the business. Transmission assets are often only half the size of generation, and they tend to expand in lockstep with new plants anyway, as mentioned earlier. Distribution systems have large asset values, often as large as total generating plants, but they grow slowly and depreciate slowly, throwing off comparatively little cash.

Data on the value of traditional IOU investments during the decade preceding 2008 bear this out. During this period the net value of all transmission assets actually *declined* by about $50 million, indicating that the total transmission plant depreciated faster than it was replaced, much less upgraded. Even more surprisingly, the net distribution plant also declined by a slight amount. In contrast, total generating capacity in this period increased by almost 20%—although essentially all of this asset growth was from *independent power producers* (IPPs).

One remarkable aspect of the coming transformation is that it will turn this thinking on its ear. My 2008 study of investment needs through 2030 for the entire power industry found that all three of the segments would require enormous new capital outlays to decarbonize generation, implement the Smart Grid, and continue to keep the lights on. We estimated that generation outlays would be $500 to $950 billion, new transmission would cost $280 billion, and distribution outlays would be the same magnitude as generation, $580 billion. The distribution figures were without the costs unique to the Smart Grid IT, whose costs we did not estimate. One anecdotal example of the high expected growth in transmission and distribution outlays comes from National Grid, the largest utility in the United States devoted exclusively to these two segments. The company expects to more than double its investment outlays during the four-year period from 2008 to 2012.[9]

These figures indicate all three segments of the industry are positioned for substantial asset growth. From the standpoint of investors, the question will be which one earns the highest expected return, taking the riskiness of the investment into account. Similarly, will combining several segments diversify risk without lowering return in a manner other securities portfolios can't match?

The answers to these questions will ultimately depend on the nature of the risks borne by pure play companies, from regulators as well as from other market and technology sources. It is a safe bet, though, that the level of technology and cost risks in the generation segment dwarf those in transmission and distribution. Remember Jim Jura's billion dollar bet? Generation planners in the coming era will have to guess whether and how fast demand in their area goes up, whether their low carbon technology works, whether their fuel will remain economical, and whether transmission will be available. All the while, DG and dynamic pricing will be playing out downstream, possibly stealing away the market for new generation entirely in one case or another.

The risks to returns on transmission and distribution assets will be plentiful, but they are likely to be smaller and involve smaller specific assets. Regulatory risks may also be lowered by national policies that encourage transmission and distribution upgrades. As an example, the FERC has already issued a policy

statement reassuring transmission owners that if they choose to prematurely retire existing transmission controls to install better smarter ones they will not
have to write off the older hardware.[10] Only time will tell how each of the segments performs as an investment, and also whether integrated utilities who reap
vertical synergies will outperform future pure plays. From the basics, however, it
does seem that the distribution-centered Smart Integrator should provide a
good investment growth story if it performs well for customers and regulators.

Who Owns the Customer?

One interesting question about the Smart Integrator model is whether it will
improve the bond between the utility and its customer. For many years, utilities'
market research has repeatedly found that utility customers feel that they have
no connection or contact with their utility. They receive a product that is invisible through a single black wire that hangs high above their home. They never see
a meter reader anymore and pay their bill monthly by mail or online. They
could easily go through their entire life not meeting a single employee of their
power company in person—all the more remarkable because they will be a customer of their utility from the moment they move into their area until the moment they leave.

The current generation of utility leaders is acutely aware of this emotional
distance and tries hard to overcome it. One glance at the annual reports or marketing campaigns of utilities today shows these attempts. A utility in my area has
a marketing slogan that reminds me that "it is connected to me by more than
wires." Progress Energy echoes this message with their slogan "Some of our
most important connections to the community aren't found on utility poles."[11]
My own utility has an ad campaign showing its linemen, fully decked out in
their tool belts and hardhats, helping serve healthy lunches and walk school-
children across busy streets.

Across every type of utility I encounter, much of the enthusiasm for the
Smart Grid arises from a belief that the customer interaction at the heart of the
new paradigm will give it a tremendously valuable means of bonding with its
customers. There is just one hitch: when you turn on your computer to set your

home's Smart Grid system, you'll feel like you're interacting with your utility only if the screen that comes up has their logo and help-line phone number on it.

Needless to say there are many other firms that sell home and business energy software that would like to provide the screen you bring up when it's time to reset your thermostat. So far, the customer interfaces in Smart Grid pilots around the country are run by utilities and so they do "own the screen." However, companies as significant as Google would like to become middlemen, and are already arguing to state regulators that utilities should have to make energy use information and control signals available to third party vendors like them. Once again, regulators will be in the driver's seat, setting the rules that allow or prevent utilities from limiting who can stand between them and their customers.

While this is a very important matter of branding and customer connection, it also bears on the core nature of the Smart Integrator business model. If utilities own the screen customers use to interact with them then the relationship is business to customer ("B2C"), whereas if the utility is serving energy management applications designers and customer premises energy management then the relationship is business to business ("B2B"). In a B2C model, the utility will need a much more extensive network of customer care specialists, much like cable companies have today, who can troubleshoot hardware and software remotely and come onsite if needed. If the model is B2B, these needs will be quite different, as its customers will be energy system vendors and installers who are infinitely more sophisticated in their interactions and needs.

Finally there is the important question of how the Smart Integrator will approach the crucial task of EE. Many utilities see EE assistance as a natural outgrowth of the energy management, control, and pricing functions provided by Smart Grid software. There are undoubtedly synergies in the provision of Smart Grid and EE investments that have barely been tapped, and that may fall naturally to the local power and information distributor.

The ultimate question here, as in the next chapter's Energy Service Utility model, is how strong the incentives are to invest in building the unique core competencies and systems needed to deliver EE, not to mention the investment dollars involved. If the EE incentives stop at decoupling, the Smart Integrator will be neutral regarding energy efficiency and will take its cues from other regulatory policies and its perception of financial and intangible benefits. All other things equal, strong incentives are better . . . which brings us to the doorstep of the Energy Service Utility.

CHAPTER THIRTEEN

The Energy Services Utility

THE MISSION of the Energy Services Utility (ESU) is to provide lowest-cost energy services to its customers—light, heat, cooling, computer-hours, and the dozens of other things we get from power each day. An ESU is a regulated entity whose prices and profits are controlled, though not without major changes to traditional cost-of-service regulation. It is responsible for supplying all retail generation customers demand with high reliability. It can own the generators that provide its supply, whether large upstream plants or small local ones, but it is also required to purchase or transmit power from others attached to its wires. Figure 13-1 illustrates the relationship between the ESU and its customers.

The ESU shares nearly all the mission elements of the Smart Integrator (SI). It must plan, expand, and operate the distribution grid reliably, including the integration of distributed generators, microgrids, and community energy systems. It must also operate essentially the same information, control, and price-setting platform the SI operates, providing universal dynamic pricing in a form approved by regulators (as in the case of the SI). As it operates this platform, it will have to contend with all the distance-to-the-customer issues we just

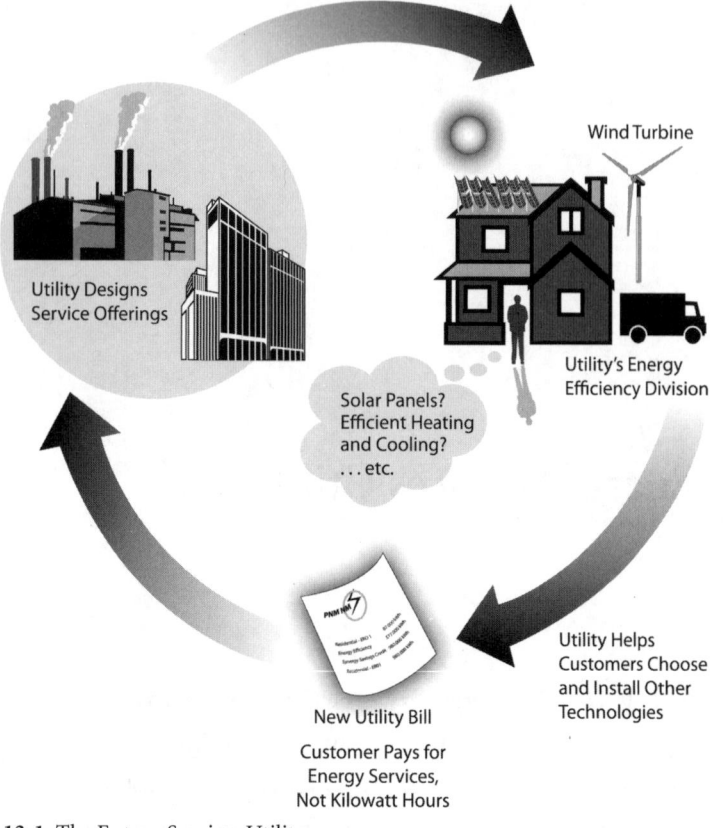

Figure 13-1. The Energy Services Utility.

examined, such as which applications and providers will be allowed onto its platform, and whether third parties can position their systems between the utility and its customer.

In the absence of retail choice and an atomistic seller-to-buyer bid/ask market, the process and software for setting local hourly prices differ slightly, though they still work fine. Every integrated utility today knows the hourly cost of the most expensive generating unit it is running at that time, and it can predict this quite accurately a day in advance. This is the correct hourly dynamic price at which the utility should offer to either buy additional supplies or sell to customers. The price has to be corrected for locational constraints using roughly the same algorithms as are used in deregulated generation markets, but that is doable too. In fact, this is just what the Pacific Northwest Labs researchers did in the Sequim experiment in Chapter 3, since utilities in that state do not of-fer retail choice.

There are two principal aspects of the ESU model that diverge from that of the SI. The first is that the ESU will not necessarily have an incentive to cooper-ate with local generators who want to connect and sell power into its smart sys-tem. Because it owns generators or contracts for supply, it may view them as the competition. The second problem is the now-familiar disincentive to help cus-tomers reduce their power use, also flowing from the ESU's ownership of gener-ation. Our ability to modify traditional regulation to overcome these two ad-verse incentives effectively will be the primary mark of success for the ESU business model.

Integrating Local Generators

There is little question that distributed generation (DG) at both the medium and the household scale is going to grow steadily over the next fifty years. Amer-ican Electric Power, a large midwestern utility, reports that DG is doubling every year on its system.[1] Even with its greater expense, customer-sited power is going to be popular among green consumers who can afford it and will be promoted by progressive state and local governments. Medium-scale power is likely to be even more numerically significant, eventually becoming the dominant sup-ply scale, some attached to the distribution system and some on high-voltage transmission.

As a regulated firm, the ESU's incentive to be cooperative in fulfilling these duties will be determined by regulatory policies. As noted in Chapter 8, the fed-eral government imposed an obligation on distribution utilities to purchase from certain small generators way back in 1978, and directed state public utility commissions to pass rules to facilitate these purchases.

By most accounts this scheme did not work especially well. Distribution utilities saw outside generators as a threat to their market share and a gen-eral headache to deal with without any corresponding upside whatsoever. The federal law also included a pricing rule that many utilities felt was a subsidy to small generators at their expense. Most of all, the scheme was adopted without any change to the fundamental incentive structure or mission of regulated distributors.

This time around these changes are an essential element of the ESU business model. The ESU must be regulated and incentivized to run the most open platform possible for new sources, in addition to its incentives to offer better pricing and help customers save energy. It is worth noting that this is the same ordeal regulators face in the SI model, where the distributor they regulate must be rewarded for planning, investing, and operating its system efficiently to integrate local supplies. The sole difference is that the ESU owns some generators, so that it may feel less willing to cooperate without stronger regulatory penalties and rewards.

A Profit Center or a Chore?

Today, utility energy efficiency (EE) programs are regulated via a ponderous process that does not promote high performance. First, regulators or public managers set out a savings goal or they adopt a planning principle that leads to a calculated savings goal. They also often specify the amount of price increase they are willing to tolerate as a result of all EE programs. Next the utility prepares detailed plans for each of its programs, including such things as the proposed level of rebates to be paid, how savings will be measured, eligible technologies, and total budgeted outlays. Regulators or the board of a public/cooperative approve the plans and budgets, and the programs begin. After the programs run for a year or so, the utility does an evaluation of savings achieved, customer satisfaction, and money spent. If the utility earns performance incentives, they are determined and allowed at this point.

The length of this cycle and the heavy degree of oversight make utility EE efforts slow-moving and highly risk-averse. Most utilities do everything they can to keep their initial savings goals quite modest so they can be sure to meet or exceed them and thereby avoid public embarrassment. Once goals are set, they want to use the most tried-and-true programs to achieve their goals, again to avoid criticism. There is no inherent reward for innovating or taking risks.

When a new idea comes up, the utility usually has to wait for another multiyear planning cycle to try it out.

As a result of all this, many utilities currently see their EE activities as something they must do to keep their regulators happy—nothing more than a mandated cost center. Decoupling does practically nothing to change this, since profits are the same regardless of their efforts. One utility CEO I know calls this a "compliance mindset." He recalls being questioned by a state regulator when he proposed being allowed to earn profits on EE. "Why shouldn't I just order you to do it?" the regulator asked him. "If you want every *i* dotted and every *t* crossed in your energy efficiency program, you will get that," he replied. "But why would any utility go beyond the bare minimum of complying with your order?"

In a few states, regulation has taken a giant step forward recently by creating genuine profit incentives for achieving high levels of customer energy savings. In California, utilities get to keep, as profit, about 12% of the value of the energy they help their customers save. In 2008, the San Francisco–based Pacific Gas & Electric Company (PG&E) earned $41.9 million profit this way.[2] In Nevada, regulators treat expenditures on EE programs, once savings are proven, essentially the same way they treat expenditures on a new power plant—both are capital outlays on which the utility earns profits. In fact, the regulations add a small premium on investments in EE, so that EE outlays earn the highest rate of return of any investment a Nevada utility can make.

These approaches, which are typically done *in addition to* decoupling, are essential for changing the mission and culture of investor-owned utilities. They focus the attention of utility managers and create a profit motive for running good efficiency programs. For the first time, they put real money at stake. PG&E's president, Peter Darbee, tells investors that "PG&E's plans are aligned with customer needs and regulatory objectives."[3] That's code for saying that the public goal of saving energy is aligned with PG&E's business interests.

EE profit incentives bring us to the threshold of the ESU but are not quite enough to unlock the door. There is still a ponderous planning and evaluation cycle, and the very real prospect of regulatory punishments, hence a lingering compliance mindset. And while they are certainly significant, the profits from efficiency are dwarfed by the profits on utilities' other lines of business. At PG&E, for example, their new investments in distribution wires, transmission, and new supplies together total about $3.3 billion, netting their shareholders about $340 million in pretax profits.

The states with utility efficiency incentives set the stage for an even more ambitious vision of the ESU. To see the emerging vision in action, we need to take two trips—one to Charlotte, North Carolina, and the second to the city of Austin, Texas.

Duke—and Others—Break the Mold

A tradition-bound utility based in North Carolina run by the same CEO for two decades is an unlikely place to incubate a new utility business model. But it is in Charlotte that Duke Energy CEO Jim Rogers has stubbornly wrestled with his own company's legacy and regulators in four states to create a business in which Duke and its profits are deeply linked to its customers' energy use decisions.

Simply put, Rogers' vision was to "redefine the boundaries" of the energy utility. In his 2008 annual report, Rogers wrote the following:

> The mission of electric utilities 100 years ago was to ensure universal access to electricity for all Americans. With that mission accomplished, the industry's mission for the 21st century is to go beyond the meter to provide universal access to energy efficiency. We must provide energy that is affordable, reliable and increasingly clean. This will drive economic growth and preserve our environment. This requires new ways of thinking about our business. . . . Our mission for this century is to redefine our boundaries—to go beyond the meter, creating new customer partnerships and providing universal access to clean and efficient energy.

In practical terms, what Rogers was talking about is the utility actually making investments in more efficient hardware that would sit in its customers' homes and businesses. As in Nevada's new efficiency incentives, these downstream investments would earn the utility regulated profits.

To move from vision to business model, Rogers and his team devised a new approach known as the Save-a-Watt program. The original proposal had three core ideas. First, the utility would make the installation of a basic package of efficiency measures an automatic part of electric service. Every home and office served by a utility has an electric meter, installed and paid for by the utility; the costs of doing this are embedded in each electric service tariff. Why not do this for a basic package of EE measures, like weather-stripping and attic insulation?

The second element of Save-a-Watt was the profit incentive. In the original proposal, regulators would allow Duke to earn as profit 90% of the difference between the costs of EE and the costs of building the next power plant, all expressed on a per-kilowatt-hour basis. For example, if financing and installing attic insulation cost an average of 3 cents/kWh saved, and a new power plant cost 6 cents/kWh produced, Duke could keep 90% of the 3 cent difference, or about 2.7 cents, as profit.

The final element of the original Save-a-Watt was an end to the heavily regulated planning and approval cycle. Since the utility was given the profit incentive to seek out and make money on EE measures, why subject it to the long, arduous, inflexible process of program preapproval? Instead, regulators would oversee independent evaluations of the true savings achieved and their true cost and allow the utility to earn 90% of the verified difference.

Save-a-Watt created quite a buzz in utility and EE policy circles. Tom Friedman praised the idea in a *New York Times* column. National efficiency advocates and environmental groups issued press releases praising the idea, and utility policymakers across the country started paying close attention to it. Rogers was profiled in the *New York Times*, *Fortune Magazine*, *Power & Energy Magazine*, *Corporate Leader*, and *Forbes* and was named one of "The 50 Most Powerful People in the World" by *Newsweek*.[4]

But back in the states where Duke needed its regulatory approvals, Save-a-Watt was in for some rough sledding. First, regulators were not comfortable giving up all control over the monies spent by Duke on ratepayers' behalf. This problem bedevils all utility EE efforts. Because regulators are often punished by their citizens or their bosses (governors and legislatures) for allowing mistakes to be made, regulators are enormously adverse to loosening their oversight of utility expenditures of any type.

Regulators also discovered that the proposed profit incentive was, in their opinion, too lucrative for the utility. To see this concern, note that the difference between very inexpensive EE measures (costing, say, 1 cent/kWh saved) and the cost of a new power plant (6 cents) is 5 cents. Under its proposal, Duke would stand to earn 4.5 cents (90% of 5 cents) on an investment of a penny, a return on investment of about 450%. Meanwhile, Duke's ratepayers would have to pay, in effect, 5.5 cents/kWh of service received, in effect paying 5.5 cents for a saved kWh that cost Duke only a penny.

Through negotiations with stakeholders and regulators, Rogers and his EE executive, Ted Schulz, have gained approval to implement their idea in four states. While regulators have insisted on significantly more oversight than was originally proposed, and have reset the profitability of the programs to levels they are comfortable with, the core elements of the idea remain. The business climate for the efficiency programs is entrepreneurial, with a premium on speed to market and maximum market penetration, not meeting arbitrarily set goals.

Ultimately, Rogers sees his utility representatives deeply involved in customer EE investments, providing expert guidance and investment funding. Much like many cutting-edge private Smart Grid energy management consultants, Rogers sees his staff customizing each one of his customers' energy management software systems to conform to the appliances they own, their comfort preferences, and control features they value. Ultimately, the ESU may even charge customers not in units of kilowatt-hours, but rather in units of heat, light, or other service provided, a concept Rogers calls "value billing."

As revolutionary as this concept sounds, a surprising number of traditional utility executives are starting to speak in the same terms. Bill Post, chairman of Arizona Public Service, the state's largest utility, says that "Factually, we provide electrons to our customers, but I don't see our business that way," he said in a recent interview. Instead, he sees himself running a service company that "cools your home, runs your computer, heats your stove," and literally powers modern-day life.[5] Ralph Izzo, head of New Jersey's Public Service Electric and Gas, says; "Just as we made universal access to energy a reality in the 20th century, so too, we can make universal access to energy efficiency a reality in the 21st. But to do this, a new business and regulatory model is needed—one that encourages utilities to invest in efficiency, in the same way we invest in pipes and wires to maintain the energy delivery infrastructure, or the boilers and turbines of power plants."[6]

"I would rather invest $10 billion in making my customers more productive with their use of energy than put $10 billion into a new nuclear plant," Rogers recently told me. If Save-a-Watt delivers on its promise, he may have his chance.

The Pecan Street Project

If Charlotte is an unusual locale for the first shareholder-based ESU, Texas is an even less likely place for a pioneering attempt to apply the model to public power. Energy production is the state's largest product; Houston is the nation's energy capital. Unlike nearly every other state, Texas remains highly committed to keeping electric deregulation for customers served by its IOUs. In an era of plug-in hybrids and smart growth, it is building a massive new intrastate highway system.

None of this bothers Roger Duncan, the manager of the City-owned Austin Energy (AE) or the city fathers of Austin, who oversee the utility's operations. AE, whose motto is "More than Electricity," is in the midst of redesigning its entire utility to be an ESU. Calling their new design Smart Grid 2.0, Austin explains:

> It[s] focus is all about the grid beyond the meter and into the premise (e.g., home, office, store, mall, building) with integration back to our utility grid. Our Smart Grid 2.0 is about managing and leveraging Distributed Generation (Solar PV, Micro Wind, etc.), Storage, Plug-In Hybrid Vehicles, Electric Vehicles and Smart Appliances on the customer side of the meter. The vision . . . is to solve the energy problem in Austin, Texas, by reinventing the power sector via moving into new energy models, including interconnecting with the transportation sector.[7]

Much of Smart Grid 2.0 involves the technologies and services that define a public power version of the SI. However, there is no retail choice in Austin; its power rates are set by the City, acting as AE's regulator.

Austin's EE offerings, though traditional in some ways, rival those of any utility in the United States. AE gives up to $1,575 in rebates for home efficiency measures and also offers assistance financing efficient air conditioners, solar photovoltaic cells, and solar water heaters. Businesses can get rebates up to $100,000 for energy management systems, $200,000 for improving data centers, and other rebates for running buildings more efficiently, curtailing loads, or installing thermal storage. Beyond rebates, Austin offers one of the pioneering municipal energy financing programs described in Chapter 10.

Most interestingly, Austin's model also demonstrates that the close political coupling between a municipal utility and local authorities can help create an integrated package of EE laws, incentives, and options. In addition to activities typically associated with utilities, Austin's program includes elements of smart growth, such as promoting alternatives to auto travel and intelligent transport systems, the creation of energy business incubators, and green workforce development. The City also adopted an ordinance that requires all single-family homeowners to perform a certified energy audit before selling their house. The audit must be given to all prospective purchasers as well as the utility. This ordinance breaks through a market barrier that has bedeviled residential efficiency investments for decades, namely the inability to demonstrate the value of an EE investment to the building's next buyer.

While all of this shows tremendous promise, Austin's transformation is a work in progress. Funding constraints have delayed the roll-out of many of the key investments, so many features of the new model are not yet up and running. It will be years before Austin will be able to gauge the full economic feasibility and degree of customer acceptance of many parts of its vision. Until then, although the vision is clearly in place, the implementation jury is still out.

The Value Proposition

The ESU is a radical concept that puts electric utilities into two diametrically opposed businesses, one selling their traditional product and one helping customers buy *less* of it. It combines the old product and service model with selling investments and services inside customer premises. Yet utilities have limited experience selling on the "customer side of the meter" and most of what they have sold to customers other than power has occurred under the compliance mind-set. They will face competition from all sorts of more nimble rivals and will be second-guessed by regulators every step of the way. What makes anyone think this business model is manageable or profitable?

One can visualize a CEO pitching the attractiveness of the ESU model to a group of skeptical, industry-savvy Wall Street analysts. What will it take for them to rate the stock of an ESU a "buy"? A utility with the capabilities to deliver customer energy services better than its unregulated competitors and earn a decent, regulator-approved profit on it. A utility able to plan and manage its generation investments so that they're located in the right parts of its system (upstream or downstream) and just large enough to backfill the residual need for low-carbon supplies. On top of all this, the utility has to do a good job managing the Smart Grid, conquering essentially all of the same challenges awaiting SIs.

The EE profit incentives conveyed by regulation are clearly the sine qua non of the ESU. As we saw earlier in this chapter, serious incentives can be grafted onto the current regulatory framework in several ways—by sharing the savings (California), treating efficiency as utility capital (Nevada), or using Duke's Save-a-Watt approach.

These may be the best practical options, but as a financial and regulatory matter they still define the utility's fiscal mission as the sale of kilowatt-hours. Every financial statement and every regulatory filing denominates the utility's accomplishments in units of energy, not units of energy saved or energy services provided.

Yet it has always been the case that customers don't purchase invisible commodity energy just for the sake of having energy. No one opens a bag of newly purchased kWh to start munching on them. Instead, customers buy power to combine it with machines that yield a cornucopia of services—light, heat, industrial motion, computing power, and hundreds of other applications. Electricity's extraordinary value stems from its astonishing versatility and efficiency at creating so many service streams from a single wire.

If it is energy services that we really want, why not measure and regulate utilities in these units rather than kilowatt-hours? Suppose, for example, that instead of setting a maximum rate utilities could charge per kilowatt-hour we allowed them a maximum rate per lumen of light delivered, including the cost of the bulb and fixture along with its input power? Under this approach, if a more efficient light source came along, so that the bulb and input power were cheaper for the utility to provide together than the existing, less-efficient combination, the utility would automatically have the incentive to install the more efficient technology.

As a simple hypothetical example, suppose that light from a typical fixture cost $3 per million lumens of light. The utility sees that a new fixture could produce the same light for $2.80 per million lumens. It would change out the lights (with the customer's permission, of course) and split the savings with the customer. Over time, as more and more customers reduced their lighting costs to $2.80, regulators would readjust the base rate, to $2.80/mm lumens, giving the utility a fair return on its investment. This is roughly how we regulated new utility power plant investments for seventy-five years. Each new power plant generated power more efficiently and cheaply than the prior generation, and regulators gradually *reduced* electricity prices over this period.

Defining utilities' mission as delivering energy services rather than commodity power is not a new idea. All of the earliest electricity vendors, including Edison, began by selling light, not power. Chicago's first lighting salesman, P.S. Kingsland, rented fifty arc lamps to customers for 15 cents an hour, while Edison's own companies competed with gas light by offering service by the bulb.[8] Beyond the pricing model, however, power companies—though not yet regulated—were deeply involved in delivering energy services, not kilowatt-hours. According to historian John Wasik:

> Power company workers in 1882 were ombudsmen. Edison's companies not only produced and sold power, they made light bulbs, all of the fixtures, wired the buildings and installed the infrastructure to generate and transport electricity. Imagine Microsoft, the software company, in addition to writing computer operating systems, making every component of the computer and supplying the electricity to run them, wiring houses and owning and running power plants. Such was the nascent electrical industry before the turn of the century.[9]

Amory Lovins and Roger Sant, among others, tried with little success to revive the idea of selling energy service in the 1980s.[10] Until recently, however, the idea seemed attractive but impractical. You cannot sell what you cannot measure, and our sensing and measuring technologies were too primitive and expensive to quantify the many different types of energy services that come from electricity. Think of the dozens of energy services electricity provides in your home: air conditioning, cooking, television, phones, computers, baby monitors . . . the list is nearly endless. How could we possibly measure, much less put a price on, the exploding menagerie of services such a prolific energy source provides?

It is unlikely that we will. However, the technology of the Smart Grid will enable us to measure some of the most important kinds of energy services with more than enough precision to set regulated prices for them if we wish. It is already the case, for example, that commercial building leases include provisions that promise to provide heat to every office within a certain temperature band—a promise that is easy to implement with readily available temperature sensors and energy management systems. Similarly, lighting and daylighting companies now contract with commercial customers to provide a guaranteed level and quality of lighting throughout their building, such as 50 foot-candles of light at a height of 3 feet with a color rendition index of 80 or higher.[11]

In the global search for new business models that promote sustainability the idea of turning a product into a life cycle service is gaining currency in many sectors. Forward-thinking carpet manufacturers like Interface will sell you carpeting services rather than carpets; you rent the carpet for as long as it lasts, and then Interface takes it back to recycle it. Aircraft engine manufacturers will write you a contract that sells you hours of service from their engines, not the engine itself, even though the engine is flying all over the world installed in your airplane. So many software providers sell you the use of their product rather than the software itself that the business model is now widely known as the software-as-a service (SAAS) model.

There are an enormous number of practical issues and barriers that remain before we can remold and regulate the utility industry mainly as an energy services business. For many electric applications, the value of electricity is greatly enhanced by consumers' ability to choose the particular bundle of features they want from among a huge variety of hardware manufacturers. In an energy services relationship, the customers lose their unconstrained choice of energy-using technologies. While this works for services for which features are not that visible or important, such as heating or lighting, imagine ceding your choice of television sets to your power company.

For the foreseeable future, the utility industry will have to experiment with this new model, applying it to the types of services that customers and regulators understand and can successfully price. If the model works well and Smart Grid technology continues to improve, service provision may eventually supplant much of the commodity sales model. Many vendors in many sectors have a business model that includes both product- and service-type offerings, and utilities will probably be no exception.

Even among the pioneers like Duke and Austin Energy, we are a long way from proving this model works. But this is the business model that must be carefully considered if we are to tap our efficiency resource without government playing the primary financing and delivery role and without retail choice. It is the only obvious policy path for states that do not want to deregulate retail power prices but nevertheless want the most advanced utility sector possible. Interestingly, it is the future business model most often mentioned by industry CEOs, even those operating in deregulated states.[12]

For public power, the value proposition obviously does not revolve directly around the rate of return on invested capital. However, the ability to deliver and realize the value within public power's constraints still must be established. Austin Energy's citizen-customers must be willing to pay as much as it costs to provide EE services as well as paying for the rest of the SI services. The City has to be able to raise and service the capital needed for energy services, keep the rest of its business running well, and keep costs and rates acceptable. Even in Texas, that's a tall order.

Conclusion

OUR JOURNEY has taken us from electricity's first revolution to its second; from Muncie, Indiana, to Austin, Texas, and Charlotte, North Carolina; from Insull's gospel of consumption to Rogers's gospel of productivity. Along the way, we've seen that changes in the technology and architecture of the grid are slowly but steadily dismantling Insull's model for the electric utility. But the journey is far from complete, and we will soon cross some dramatic terrain.

The Smart Grid will give customers much greater control over their power use and make dynamic pricing universal in one form or another. It will force extensive changes in utilities' operating procedures and unleash a tidal wave of new regulatory challenges. Much of the regulatory conflict will center on the perennially difficult issues of investment benefit measurement, allocation of systemic costs, and the process of blending markets and regulation—problems that cannot be solved by simply "letting the market decide."

Electric sales will grow slowly, if at all, for the next few decades. Nevertheless, the need to scale back carbon emissions calls for hundreds of thousands of megawatts of new generation, much of it from sources that are commercially unproven, expensive, or difficult to tap. Distributed generators will provide a steadily growing share of this power, but their costs will remain well above most large-scale plants and their growth will be limited by the rate at which the grid can be reengineered.

Despite certain inconvenient attributes, we will build many large natural gas, wind, and solar plants and even more medium-sized versions of the same. Coal plants with carbon capture or the next generation of nuclear units may prove economical enough to be part of the mix by the 2020s, especially considering the large international R&D efforts behind them, global energy geopolitics, and the domestic political support for these two forms of energy.

Financing this program of decarbonization on a stagnant sales base will be challenging enough, but many new generators will also need transmission additions. The 20,000 to 30,000 miles of new lines needed falls far short of a nationwide transmission supergrid, but necessitates decision making and construction at a much faster pace than the industry has achieved in modern times. The acceleration will require new methods of regional planning and grid cost allocation that force a truce in a grid policy war that has gone on for decades.

Most important of all, the cardinal role of energy efficiency in any serious national climate strategy requires policy choices and regulatory reform. Properly decoupling sales from profits is a useful first step, but it will fall short of the metamorphosis needed for the coming era. The industry needs a new form of regulation and new business models to match its new technologies and mission.

Second-Century Regulation

At its core, economic regulation is an instrument that rewards private firms for shifting from the actions they would choose if left unfettered to efforts more in line with our economic and social goals. The policymakers who fashioned electric regulation in the 1920s and 1930s understood this. They designed regulation to do what America then wanted: the largest, cheapest commodity power system possible, ignoring most environmental constraints and the promotion of energy efficiency. In rural areas, where investor-owned firms could not invest profitably, government-financed cooperatives would take their place; cities that had municipal utilities with largely the same objectives would also grow. Cheap, universal commodity power served a national mission to build a world-class infrastructure and increase our productivity to the point where our security was assured and our middle class became the wealthiest consumer society in history.

These goals remain, but they are now joined by a realization that the world's carbon budget is vastly overspent and our economic infrastructure was not designed for sustainability. A larger electric power industry is no longer *automatically* the route to greater productivity and a better standard of living. In the coming century the power industry we seek is not necessarily the largest, it is one that can help its customers achieve the highest level of service possible consistent with social and environmental sustainability. Its goal is not more, but more from less.

The boundaries and business strategies of the utilities of the future will be set by the objectives and incentives regulation provides. There are two potentially feasible paths. In states that adopt or continue retail choice, Smart Integrators can be regulated to operate energy and information platforms accessible to all consumers and suppliers on a policy-consistent basis. Where conditions allow, the interplay of demand and supply can set prices, but the role of regulation establishing and protecting such markets, and overseeing the distribution utility at the hub of the network, will create generations of work for regulators.

The second path leads from today's integrated utilities, who continue to supply commodity or provider of last resort (POLR) power. These utilities must operate the same efficient, policy-consistent delivery platforms, but their incentives to favor their own generation supplies bespeak a need for stronger energy efficiency incentives and open access policies. Here regulation needs surgery much more extensive than the grafting of decoupling onto the cost-of-service framework. Energy Service Utilities must be regulated so as to be rewarded for delivering lowest-cost energy services, whatever the combination of generation, delivery, and efficiency capital needed.

New Roles and Resources

Developing a new regulatory compact to fit the industry's new mission is a Herculean task. Conceptually, the goal is to create rate-setting formulae that create the incentive to invest efficiently for lowest-cost energy services while meeting reliability rules, carbon limits, renewable standards, and other constraints. Experience with other forms of incentive regulation shows that this will require experimentation, cutting-edge regulatory skills, and no small measure of fortitude.

Absent very unlikely changes in federal law, this task will fall to fifty state legislatures, governors, and utility commissions. Most legislatures and executive branches will need to devote a huge effort to understanding the industry's changes and their options for reform. They must also avoid being captured by one or another stakeholder, or simply taking the political path of least resistance regardless of its ultimate viability. State commissions are in a little better shape, having accumulated valuable experience and developed effective institutions for sharing information and improving their quality. But little in their training or history has prepared them for a mission so sweeping as the redesign of regulation itself.

Today the average regulatory commission has a staff structure and core competencies befitting its historic mission. It is composed primarily of lawyers and accountants, many of whom have acquired an excellent understanding of public utility economics and electric technology. There are a few engineers, even fewer economists, and an extremely limited number of experts in electric information technology.

These agencies do amazingly well with their limited resources, but recent history has not been especially kind to them. Prior to the upheaval of the late 1970s, commissions were sometimes considered state government backwaters that carried out nothing more than routine accounting tasks. In the late '70s, inflation, twin energy shocks, and the cost overruns from early nuclear plants sent electric rates to sky-high levels. State leaders lost faith in utility regulation as a means of controlling excessive prices, and deregulation became the rage. Utility commissions were caricatured as relics of a bygone era, with little or no future role in the coming free energy markets. Budgets were cut, the best and brightest staff members left, and a commissioner job became the least sought-after spot for an ambitious up-and-comer.

As states gained experience with retail choice, they gradually learned that their regulators had a much greater responsibility overseeing the system than they originally understood. "Deregulation" required state regulatory commissions that needed just as much authority, expertise, and independence as in regulation's heyday, along with many new skills needed to understand complex new markets. Meanwhile, when prices spiked or expectations were not met, elected leaders in the state needed to blame someone. State utility commissions were often the obvious choice.

Regulation by an independent commission is founded on a compact in which regulators act quasi-judicially, without direct political interference by either the governor or the legislative branch.[1] Unfortunately, the California energy crisis and the perceived failure of retail deregulation has ushered in an era in which legislatures and governors have bitterly criticized, threatened, and sometimes replaced public service commissioners for alleged failures to protect the public. After two decades of frequent criticism, some utility commissions are understandably resistant to bold new regulatory experiments and reluctant to allow utility rates to increase.[2]

Nothing could be more harmful to the future of the industry than a state regulatory community too disempowered to manage professionally and impartially the profound changes needed in regulation today. It is a matter of national importance that state commissions have their independence restored and their missions and resources reset to play their part in a transformation that is already under way and will not stop.[3] States need not change their policies on retail choices, but they must revise their state regulatory compacts and upgrade their commissions if they hope to reap the full economic benefits of the Smart Grid and turn greenhouse gas policies into an engine of economic development.

A complete exploration of the policy and management changes state commissions deserve is a study in itself, if not a library of studies. However, a few suggestions along these lines seem both obvious and feasible. The federal government should dedicate funds sufficient to establish regulator training programs that rival the best academic and professional programs in the world. The regulatory community should use these funds to establish a series of professional accreditation programs. The university of modern regulation should have courses of study in electric pricing methods, market oversight skills, communication networks, and other specializations.

Without constraining state policies for appointing regulators or hiring senior staff, state commissioners and senior staff members should be required to gain accreditation following their confirmation. The accreditation program should also be available to federal regulators, governors' energy staffs, and the leaders of public power agencies and cooperatives. This educational requirement is as much a means of imbuing the culture of regulatory commissions with an appetite for embracing and managing change as it is a means of imparting specific knowledge.

In many states regulatory commissions have a number of responsibilities beyond electric regulation and are hard-pressed to carry out all their duties with their current resources and staff. The level of staffing varies widely in commissions, from thirty in Mississippi (2 for every 100,000 ratepayers) to 996 in California (if you count both the California Public Utility Commission and the California Energy Commission, this is 7 for every 100,000 ratepayers). To the maximum extent possible, the electric and communications regulatory functions should be given their own organizational structures, ample budgets, and the resources and authority to hire excellent new staff where needed.

Commissions in many states (as well as the FERC) have made great strides toward changing the process of regulation away from expensive and often ineffective adversarial litigation to collaborative, technical conferences, and other modern processes. The new processes are ideal for examining the changes needed in the regulatory compact, the benefits and costs of Smart Grid investments, energy efficiency policies, and many other new challenges. Their use should be encouraged and expanded.

An upgrading of utility commissions bears some similarities to the resurgence of interest in revamping financial regulation in the wake of the 2008 market crash. In both cases, regulatory agencies and their rules were created for industries with the technologies, products, and business operations of the early twentieth century. In both cases, the industries have changed so much that the old regulatory processes and goals no longer serve our national objectives. In both cases, the magnitude of the challenge commands us to create the smartest, most agile, and most highly skilled regulatory agencies possible.

Public Power's Moment of Opportunity

Many of the forces buffeting investor-owned utilities (IOUs) are affecting municipal utilities and cooperatives with equal force. All segments of the industry face an imperative to adopt the Smart Grid cost-effectively and allow greater customer control via dynamic pricing. As we saw with Jim Jura and Associated Electric, they also face the same choices and uncertainties in supply options and they are confronted by a similar need to meet climate change limitations.[4]

These imperatives give public power and cooperatives the same challenges that IOUs must contend with, revising their mission and management and acquiring new core competencies. If anything, the introduction of dynamic pricing and the other core elements of the Smart Grid will be a bigger change in operations and culture for many utilities in these segments than it is for most IOUs. With smaller staffs and resources, publics and cooperatives will have the equally difficult task of specifying, procuring, and operating IT platforms that are still evolving rapidly, subject to rapid obsolescence, and uncertain as to their acceptance by customers. Unlike IOUs, which keep a large layer of shareholder equity, utilities in these segments have little protection against failed generation or distribution investments. Yet the coming era is one of unprecedented new risks in every kind of utility outlay.

In these segments, there is less of a distinction between government and utility leadership in energy efficiency because the utility is itself quasi-governmental. The challenge is rather the means of inducing these utilities to make large customer efficiency investments. There is no external regulator that can reward shareholders for doing a good job via monetary incentives, there are only city councils and cooperative boards. These leaders can praise managers that do a good job saving energy or even pay them bonuses, but this is largely unexplored territory.

Because most public power and cooperatives have steered clear of retail choice thus far, the obvious model for them is the Energy Services Utility.

However, in addition to a lack of external efficiency incentives, heavy investments in upstream supplies and the lesser ability to absorb losses on investments already made will create especially strong conflicts between saving and selling more power.[5] All in all, aligning the incentives of the managers of public power firms with the national imperative of making customers more electricity-efficient is an enormous and unresolved challenge.

These changes paint a picture of adversity for the public and cooperative segments every bit as large and messy as the landscape confronting IOUs. At the same time, there is an aspect of the coming changes that presents these segments with one of the largest opportunities in their history. The current structure and architecture of the industry, with very large generators at the far end of the high-voltage grid, favors large utilities operating across enormous territories. The coming age of decentralized sources and a smart grid will favor community-scale sources located much closer to load.

This is an industry configuration that is inherently matched to municipal utilities and cooperatives. These utilities are effectively owned by their communities, typically serving areas with medium-sized electric demands and geographic footprints. Community energy systems and so-called microgrids are well-suited to these ownership forms.[6]

Historically, the IOUs have been very resistant to any expansion of public power. A growth of community energy systems in Smart Integrator environments will require leadership and investment at a time when state and local governments are besieged by problems from the economic downturn that began in 2008. In addition to possible IOU opposition, a hungry new generation of green power entrepreneurs would like nothing more than to own and profit from the microgrids and community energy systems; many IOUs also have deregulated subsidiaries that are already active.

Nonetheless, the municipal and cooperative segments are facing a unique, back-to-the-future moment. The industry's architecture is being redesigned to operate on a scale that better suits these segments, giving them an opportunity to own and manage supply resources at the scale they have dominated for years. In Muncie, Indiana, the small municipal utility, integrated into the massive investor-owned American Electric Power System in 1911, may someday in the not-too-distant future host a community energy system owned by the city or another public agency.[7]

Making Power Smarter

Regardless of what we do, the lights in America will stay on. Barring an unusually successful physical or cyber attack, the grid will remain reliable. Electricity will remain an absolute necessity, and most families and businesses will find it affordable. State regulators will allow utilities to add greater intelligence to their distribution systems and evolve toward dynamic pricing. Following enactment of nationwide carbon emission limits, utilities will find ways of reducing their greenhouse gases.

The question is not whether we make these changes, but whether we make them well or poorly, costly or cost-effectively, quickly or at a tortured, halting pace. Mother Nature's timetable for the safe decarbonization of the power sector is not negotiable. With outright deregulation far too unpopular to offer a simple fix, an aging regulatory approach based on deficient public objectives must be changed firmly and quickly, with parallel changes in the public power and cooperative segments led by the sectors themselves.

Much of the art of good public policy is learning when to let the marketplace, and the marketplace of ideas, run its course. It is unnecessary and probably counterproductive to force a choice between the Smart Integrator model and the Energy Service Utility. As decarbonization proceeds and the Smart Grid inhabits more of the industry's skeleton, there is no doubt that the better model will reveal itself if one is, in fact, better. It is necessary only that each state be clear about its model and work hard to regulate it wisely.

There is, however, a critical decision we must make as a nation, and soon. We must choose between putting state and local governments in charge of financing and delivering energy efficiency or making it part of the industry's mission and business. More accurate electricity and carbon prices are essential, but not nearly sufficient to overcome the full range of market barriers. A source of patient capital and trusted expertise is certainly needed. And the power of the purse must be wielded, and wielded wisely. As the old saying goes, if everyone is in charge, no one is in charge. We cannot expect efficiency to deliver its store-house of low-cost savings without a motivated and accountable quartermaster.

Few of the changes I advocate need to come from new federal laws. National policy clarity is needed on leadership in energy efficiency, and national resources are needed to upgrade state commissions. Beyond this, the main locus of institutional and regulatory change will occur in states, localities, and multi-state planning regions, where governors must find new ways to cooperate on electricity grid expansion.

As Insull knew, it was the cohesion between the industry's economic properties, its division into firms and markets, and the incentives regulation conveyed that allowed the power industry to flourish. State by state and utility by utility, the mission, structure, and governance of utilities must change. There will and should be many variations, but without a harmonization of economics, structure, and sound regulatory governance the industry will not perform well.

In the end, a smart power industry will not be the product of the oncoming revolutions in control systems or generating technologies, grand as they are. It will be the result of provisioning the industry for change. The intelligence of the institutions we create, not that of the hardware and software we deploy, will determine whether the industry that created the world's wealthiest and most powerful nation will lead that same nation to a new, more productive, and more sustainable future.

APPENDIX A

Electricity Sales Scenarios

THIS APPENDIX describes the EIA electric sales forecast and adjustments discussed in Chapter 6. The master table to this appendix, Electric Sales Scenarios is shown on the next page. I discuss each portion of the table in order from top to bottom.

Energy Information Administration Forecast

The baseline for all scenarios is the April 2009 Energy Information Administration (EIA) updated base case, which reflects the American Recovery and Reinvestment Act of 2009 (ARRA) but no other Obama-era energy policy enactments. Table A-1 shows some of the results of this EIA forecast.

In this discussion, increased economic growth serves as a useful proxy for economic growth, population and household growth (both number and physical footprint), and related variables. One could create many combined scenarios with various combinations of these key drivers, but my goal is simply to create several representative scenarios, not the full range of possibilities.

In the original Annual Energy Outlook (AEO) 2009 update, which relies

Table A-1. Selected EIA Annual Energy Outlook Results
April 2009 Update

	2007	2030	Growth (total T or AACGR)	Source*
GDP (trillion 2000 dollars)	11,524	19,875		Table A2
Electric Sales (billion kWh or TWh)	3,747	4,527		Table A8
Average Retail Electric Price				
(2007 ¢/kWh)	9.1	10.1	11% T	Table A1
Total Energy Used (quadrillion Btu)	101.90	110.95	8.8 % T	Table A2
Households (millions)	113.74	142.08		Table A4
Average House Square Footage	1,663	1,934		Table A4
PHEVs (cumulative)		4.7 MM		

Note: *Department of Energy, Energy Information Administration, "An Updated Annual Energy Outlook 2009 Reference Case Reflecting Provisions of the American Recovery and Reinvestment Act and Recent Changes in the Economic Outlook," *The Annual Energy Outlook 2009*, U.S. Department of Energy. SR-OIAF/2009-03, April 2009. PHEV figure based on related supplemental reference tables.

on IHS Global Insight, Inc.'s macroeconomic forecasts, the base, high, and low gross domestic product (GDP) average real growth rates are 2.5, 3.0, and 1.8% per year, respectively (http://www.eia.doe.gov/oiaf/aeo/assumption/macroeconomic.html). The higher and lower GDP electricity sales are generated from the linear relationship between GDP and power sales reflected in the annual results of this EIA model run.

As shown in the first block, 2030 GPD rises to 19.9, 22.1, and 17.7 trillion ($2007), respectively, in the base, high, and low growth scenarios. Electricity use is 4,527, 4,794, and 4,315 billion kWh, respectively. Note that EIA's reference forecast shows that the U.S. economy will become substantially more electric-efficient over the 23 years of the forecast—the electricity used per unit of real GDP drops 30% during this period, from .325 kWh/$07 to .228 kWh/$07. EIA's forecast should therefore be understood to include very significant price- and technology-driven energy efficiency (EE).

Plug-in Hybrids

Table A-2 shows that EIA's reference forecast reflects a much slower penetration of plug-in hybrid vehicles (PHEVs) than many other studies, perhaps because it

Table A-2. Electric Sales Scenarios

	Units	2007	2008	2030	2007–30 Growth	2007–30 AACGR
PROJECTED ELECTRIC SALES						
GDP: EIA April 09 Reference Case [1]	(2000 $Bil)	11,524		19,875	72%	2.40%
Electric Sales: EIA April 09 Reference Case [2]	(Bil kWh)	3,747	3,725	4,527	21%	0.83%
Sales/GDP	(kWh/$)	0.33		0.23	–30%	
Growth in Electric Sales, 2008–2030	(Bil kWh)			802		
High GDP Growth Rate [3]	(2000 $Bil)	11,524		22,142		2.88%
Implied Electric Sales	(Bil kWh)			4,794		
Sales/GDP	(kWh/$)			0.22		
Low GDP Growth Rate [4]	(2000 $Bil)	11,524		17,370		1.80%
Implied Electric Sales	(Bil kWh)			4,315		
Sales/GDP	(kWh/$)			0.25		
PROJECTED LIGHT-DUTY PHEVS IN USE						
EIA PHEVs in Use [5]	(000)			4,068		
Electricity Used by PHEVs [6]	(kWh/Vehicle/Yr)			1,439		
TBG High PHEVs [7]	(000)			10,657		
Electricity Used by PHEVs [8]	(kWh/Vehicle/Yr)			2,560		
Incremental Electric Sales [9]	(Bil kWh)					
from EIA fleet using higher % electric drive				5		
from incremental vehicles				17		
Additional Electric Sales from High PHEVs in Use	(Bil kWh)			21		

Table A-2. Continued

	Units	2007	2008	2030	2007–30 Growth	2007–30 AACGR
LONG-TERM PRICE IMPACTS ABOVE EIA FORECASTS						
EIA Real Price Increase 2007–2030 [10]	(%)			10%		
Assumed Higher Price Increase Scenario [11]	(%)			13%		
Long-Term Price Elasticity [12]				−0.7		
Reduction in 2030 Sales	(Bil kWh)			−86		
POLICY-DRIVEN ELECTRIC DSM EXCLUDING SMART GRID ENABLED						
RAP Case Savings [13]	(Bil kWh)			−398		
Max Case Savings [14]	(Bil kWh)			−544		
Max Savings Share of Reference Case Sales	(%)			12%		
SMART-GRID ENABLED DSM, INCLUDING ADDITIONAL PRICE RESPONSE						
Smart-Grid Savings [15]	(Bil kWh)			−181	4%	
ADDED ONSITE GENERATION DUE TO POLICY SHIFTS AND SMART GRID						
EIA April 2009 Projected Base [16]	(Bil kWh)			318	−1	
Increase Cases	EIA Waxman–Markey Simulation					
EIA 2000 Policy Scenarios [17]	Simulation			−150		
Advanced Technology				2020		
Advanced Technology Plus Net Metering				−5		
AT Plus Tax Credit				−18		
Reduction in 2030 Sales [18]				−30		
TOTAL ADJUSTMENTS AS REPORTED IN TEXT (BIL KWH)				−674		

TOTAL U.S. SALES, REFERENCE CASE NET OF ADJUSTMENTS (BIL KWH)   3,853
Difference from 2008 Actual Sales   128
% Increase from 2008 Actual Sales   3.4%

Notes:

[1]: EIA 2009 AEO (April 2009 release); SR/OIAF/2009-03; Reference case table 20.

[2]: EIA 2009 AEO (April 2009 release); SR/OIAF/2009-03; Reference case table 8.

[3]: Assumed.

[4]: Assumed.

[5]: EIA 2009 AEO (April 2009 release); SR/OIAF/2009-03; Reference case supplemental table 58.

[6]: Derived from EIA 2009 AEO (April 2009 release); SR/OIAF/2009-03; Reference case supplemental tables 47 and 58.

[7]: EIA trajectory, scaled up by goal of 1 million vehicles in 2015.

[8]: Assumes 300 watts per mile: EPRI-NRDC Environmental Assessment of Plug-In Hybrid Vehicles, vol. 1, p. 4-4 Table 4.2. Reflects 67% of 12,736 total miles per year driven with electric drive (average miles per day of 34.89 as reflected in AEO data).

[9]: 67% from EPRI-NRDC Environmental Assessment of Plug-In Hybrid Vehicles, Vol. 1 p. 5-2 Table 5.1.

[10]: EIA 2009 AEO (April 2009 release); SR/OIAF/2009-03; Reference Case Table 1.

[11]: Assumed.

[12]: Assumed.

[13]: EPRI Assessment of Achievable Potential from EE and DR Programs in the United States 2010-2030 Table 4-1.

[14]: Equals -398, scaled by 11.2%/8.2% in Figure 4-1 of EPRI Assessment.

[15]: EIA 2009 AEO (April 2009 release); SR/OIAF/2009-03; Reference Case Table 8: Total End Use Generation.

[16]: EIA 2009 AEO (April 2009 release); SR/OIAF/2009-03; Reference Case Table 8: Total End Use Generation.

[17]: Boedecker, Erin, John Cymbalsky, and Stephen Wade, "Modeling Distributed Generation in the NEMS Buildings Models," Energy Information Administration, July 30, 2002. Tables 3–10. Available at: http://www.eia.doe.gov/oiaf/analysispaper/electricity_generation.html.

[18]: Year 2020 change from reference case increased linearly from 2010-2020 increase.

Table A-3. PHEV Penetration Scenarios (Light Duty Vehicles in Thousands)

	Sales (2030)	Penetration (%)	Vehicles on Road	Penetration (%)
EIA April 2009[1]	428	2.3%	4,068	1.4%
Oak Ridge National Labs[2]	5,200	25%	50,390	
Electric Power Research Institute/Natural Resources Defense Council[3]				
Medium Scenario	9,000	50%		
Low Scenario	3,600	20%[4]		
"Base Case"	900	5%		
MIT Laboratory for Energy and the Environment[5]				
"Hybrid-Strong" Scenario	1,620	9%		
Brattle Group–Obama Administration Goals[6]	1,062	5.9%	10,657	3.6%

Notes:

1. Department of Energy, Energy Information Administration, "An Updated Annual Energy Outlook 2009 Reference Case Reflecting Provisions of the American Recovery and Reinvestment Act and Recent Changes in the Economic Outlook," *The Annual Energy Outlook 2009*, U.S. Department of Energy. SR-OIAF/2009-03, April 2009, Table 45.

2. Stanton W. Hadley and Alexandra Tsvetkova, "Potential Impacts of Plug-In Hybrid Electric Vehicles on Regional Power Generation," Oak Ridge National Laboratory, Oak Ridge, TN; for the U.S. Department of Energy, DE-AC05-00OR22725, January 2008. Sales Figure 3; Fleet Table 1; page 5.

3. Electric Power Research Institute, "Environmental Assessment of Plug-In Hybrid Electric Vehicles, Volume 1: Nationwide Greenhouse Gas Emissions," Technical Report, Electric Power Research Institute, Palo Alto, CA, July 2007. Page 6, Based on 18MM Annual Sales 2030.

4. Figure is for 2050, not 2030.

5. A. Bandivadekar, K. Bodek, L. Cheah, C. Evans, T. Groode, J. Heywood, E. Kasseris, M. Kromer, and M. Weiss, "On the Road in 2035: Reducing Transportation's Petroleum Consumption and GHG Emissions," MIT Laboratory for Energy and the Environment, Cambridge, Massachusetts, 2008. Figures 51 and 53. Sales based on 18MM total LDV sales in 2030, reference A. Available at: http://web.mit.edu/sloan-auto-lab/research/beforeh2/otr2035/

6. "Promoting Use of Plug-In Electric Vehicles through Utility Industry Acquisition and Leasing of Batteries, Chapter 13 of 'Plug-In Electric Vehicles: What Role for Washington?,'" by Peter S. Fox-Penner, Dean M. Murphy, Mariko Geronimo, and Matthew McCaffree, The Brookings Institution, 2009.

assumes no policy shifts. Even the Massachusetts Institute of Technology (MIT) report's base case (referenced in Table A-2), which also does not assume a particular policy emphasis on PHEVs, predicts nearly twice as many PHEVs sold in 2030 as does EIA.

The sales adjustment I employ is based on the final line in the table. This is a scenario developed by my colleagues Dean Murphy and Mariko Geronimo by scaling up EIA's forecast to meet President Obama's goal of one million plug-ins in the fleet by 2015. Table A-3 shows that this scenario requires a little over double the very low sales penetration in EIA's forecast, much less than the remainder of the studies surveyed, with the exception of MIT's base case. While it is easy to envision scenarios with higher or lower PHEVs, the scenario we employ is a plausible representation of an achievable policy-driven boost above EIA forecasted levels.

In addition to adjusting the number of vehicles sold, I also adjust the electricity used by each vehicle. EIA assumes that about two-thirds of the PHEVs sold in 2030 will have ten-mile ranges, while most other analysts assume that 40-mile-range vehicles will be standard. This impacts the electricity sales to vehicles substantially, as the majority of vehicle-miles traveled by drivers in PHEV-10s do not use electric power. In contrast, the more optimistic studies, such as Electric Power Research Institute/Natural Resources Defense Council (EPRI/NRDC), assume that about 67% of all miles are driven in electric mode. The difference between these assumptions results in annual sales to vehicles of 1,439 kWh in EIA's forecast and 2,560 kWh/year using EPRI/NRDC results.

To create a more aggressive scenario, I employ the EPRI/NRDC assumptions concerning sales per vehicle, though I do not employ their fleet penetration assumptions. The accompanying Table A-3 shows the two adjustments to EIA's forecast for PHEVs, one to increase penetration to the *Brattle*–Obama goals scenario and the second to increase electric use per vehicle.

Long-Term Price Impacts

Long-term price impacts are the product of real price increases over and above those in EIA's forecast and the long-term price elasticity. EIA's forecasts of price increases should not include the impacts of a carbon cap and trade bill, a renewable electric standard, or any of the policies under discussion.

A plethora of studies attempt to examine the retail impacts of recent proposed energy policies, including several specialized analyses by EIA itself. A small sample of the results of these studies is shown in Table A-4.

No single study examines the impacts of all proposals under discussion, nor

Table A-4. Retail Impacts of Recent Proposed Energy Policies ($/kWh)

Organization	2030 Electricity Prices (Baseline)	2030 Electricity Prices (with Waxman–Markey)	Percentage Increase
Environmental Protection Agency (2008$)	$0.099	$0.113	14.14%
Charles River Associates (2007$)	$0.100	$0.122	22.0%
Energy Information Administration–A (2007$)	$0.100	$0.120	19.56%
Energy Information Administration–B (2007$)	$0.101	$0.101	0.00%

Sources:

"EPA Analysis of the American Clean Energy and Security Act of 2009," U.S. Environmental Protection Agency, June 23, 2009. Based on figures from the Data Annex. Available at: http://www.epa.gov/climate change/economics/economicanalyses.html.

David Montgomery et al., "Impact on the Economy of the American Clean Energy and Security Act of 2009 (H.R.2454)," Prepared for National Black Chamber of Commerce, CRA International, May 2009. Figures in the table assume a baseline similar to that of EIA's.

(EIA-A): "Energy Market and Economic Impacts of H.R. 2454, the American Clean Energy and Security Act of 2009," Energy Information Administration, Report SR/OIAF/2009-05, August 2009.

(EIA-B): "Impacts of a 25-Percent Renewable Electricity Standard as Proposed in the American Clean Energy and Security Act Discussion Draft," Energy Information Administration, Report SR/OIAF/2009-04, April 2009.

"Ways in Which Revisions to the American Clean Energy and Security Act Change the Projected Economic Impacts of the Bill," U.S. Environmental Protection Agency, May 19, 2009.

is it likely that all proposals become law in their present proposed form. Moreover, a national climate policy is very likely to rebate a significant portion of policy-derived revenues, though not in a form that lowers marginal power prices.[1] For the purposes of this calculation, this means that the sales-reducing price impact will be slightly offset by a positive income effect, relative to a policy with no rebates.

To remain conservative, I have chosen a 3% real price increase in 2030 attributable to policies not reflected in EIA's forecasts. This is extremely close to EIA's estimate of the 2030 impact of just the renewable energy standards (RES) portion of Waxman–Markey (W-M) as estimated in April 2009, albeit for 2025 rather than 2030.[2] My estimate is lower than many other forecasts of the impact of carbon cap-and-trade legislation, but there is also a history of overestimating policy compliance costs,[3] and my intent is to be conservative in any case.

For the elasticity term I employ a long-term price elasticity of –0.7. This is approximately the level reported in Bernstein's recent research and below the level of most older estimates. Erring on the conservative side here is also directed toward avoiding double-counting of these effects with EE and Smart Grid–enabled efficiency.[4]

Additional Policy-Driven Energy Efficiency

At any price level, and with or without a carbon policy, EE efforts can be increased or decreased through additional efforts ordered or incentivized by state or federal agencies. The proposed W-M legislation contains a number of policies that would be expected to boost electricity savings above the EIA baseline, including the ability to satisfy up to 20% of the RES mandate using verified efficiency savings, accelerated building standards, and a federal program to mandate higher building efficiency. A senate energy bill just reported out of Committee also has accelerated EE provisions.

At the state level a variety of EE policies are likely to trend upward over the next twenty years, probably in areas federal policies do not emphasize. Traditionally, state EE incentives and mandates have primarily come in the form of directed levels of effort and sometimes via the ability to use EE program savings to comply with renewable portfolio standards (RPS) mandates. The American Council for an Energy Efficient Economy's (ACEEE's) comparison of state efforts shows that there is wide variation in state policies, and suggests that most states could increase their programs substantially and remain cost-effective.[5]

Measuring the impacts of additional EE policies raises all of the issues inherent in any measurement of EE potential and the cost of overcoming market barriers, topics discussed in more detail in Chapter 10. In this exercise we assume the policy and measurement issues are sufficiently settled so as to enable valid estimates of true electric sales savings incremental to the EIA reference case.

The EPRI study[6] used as the basis of my sales adjustment for expanded EE efforts does not specify a policy approach, instead focusing on the available cost-effective potential from a set of traditional (non–Smart Grid) efficiency technologies. Utility or state programs that target these measures should be able to achieve the estimated savings over and above the EIA baseline. I employ the "realistically achievable potential scenario" (RAP) results of 398 billion kWh, the more conservative of the two EPRI cases. This scenario has been criticized as being much too low by some efficiency experts;[7] to the extent that it is, the higher "max case" of 544 billion kWh is more appropriate and my conclusions regarding the possibility of flat or negative sales growth are stronger.

It is an axiom of good EE planning that these programs should not target, reward, or count as savings efficiency efforts that would have occurred anyway (so-called free riders). Accordingly, properly designed EE policies will not create savings that would have occurred in the EIA baseline and in the additional unassisted price response cause by own-price elasticity. Moreover, my use of the "realistically achievable" EE savings scenario as well as a relatively low price response adjustment is also intended to offset what is certain to be a small amount of overlap between savings categories.

A preliminary indication that my adjustment looks reasonable and achievable comes from an analysis of the impact of W-M on electric demand prepared by the ACEEE, which estimates that Titles I and II of the bill, which primarily involve the renewable energy/energy efficiency portfolio standards, accelerated building and appliance standards, and building and lighting EE, could save 505 billion kWh by 2030.[8] This is 25% more than the EPRI RAP scenario, but quite similar to the "maximum achievable" EPRI results.

Smart-Grid Enabled Energy Efficiency

There is a class of EE technologies, and their accompanying savings, that are enabled by Smart Grid technologies. The simplest such technologies are visual indicators of current hourly energy prices, often in the form of globes that change color as prices change, known as in-home displays (IHDs). Many other appliances will be able to be remotely controlled. This will allow customers to shift load, which does not reduce sales, but in some cases will allow customers to turn off unneeded end uses more easily, locally or remotely. This is certain to reduce device-driven energy use, which is the fastest growing category of power use in the United States today.[9]

Although some of these savings should be reflected in EIA's forecasts, most of them involve policy changes that will enable Smart Grid deployment. They should not be reflected in the EPRI EE potential study because the EPRI study included only existing devices and practices in its estimates. In any event, our understanding of the size of EE savings the Smart Grid will enable is very preliminary, based on a subset of the approximately 40 Smart Grid pilots under way or completed in North America.

The most complete analysis of the particular EE measures and programs specifically enabled by the Smart Grid comes from a 2008 EPRI study, *The Green Grid: Energy Savings and Carbon Emissions Reduced Enabled by a Smart Grid.*[10] The study finds five areas in which Smart Grid–enabled technologies create significant incremental savings: (1) continuous commissioning of buildings and other information-enabled technologies; (2) improved operational efficiency of the distribution system, reducing line and substation losses; (3) enhanced demand response and load control, which has a small energy savings effect; (4) reduced customer use purely from observing real-time prices; and (5) better utility EE programs simply because information on savings will become more accurate.

For each of these five areas, EPRI created three penetration scenarios as of 2030. For example, continuous commissioning was assumed to occur in somewhere between 5 and 20% of all commercial buildings by this date. Customer feedback was assumed to occur in somewhere between 25 and 75% of all American homes. Overall, EPRI concluded that Smart Grid–enabled EE technologies would save between 56 and 203 billion kWh by 2030. ACEEE has also done a recent, preliminary estimate of Smart Grid–enabled EE; ACEEE finds that nearly six times as much electric power can be saved through a "semiconductor-enabled" efficiency strategy (1,242 billion kWh).[11]

I also draw on the work of my *Brattle* colleagues, led by Ahmad Faruqui.[12] Ahmad's team observed savings in the range of 4 to 7% from residential customers purely from the effect of seeing price information, without increased ability to control devices. One would expect that device control, when coupled with price transparency, would increase savings significantly, since it reduces the cost of the efforts customers evidently are putting out to reduce their use manually in response to prices.

Offsetting these considerations, it is highly unlikely that the Smart Grid will have penetrated all service areas, including rural utilities and difficult-to-reach customer segments. Because Smart Grid deployment is highly dependent on the pace of standardization, technological change, and regulatory policies, there are very few analytically based deployment forecasts. One can think of the estimate as a combination of 50% deployment of the Smart Grid and 8% savings from Smart Grid–enabled EE or other similar combinations yielding approximately the same outcome.

To avoid double counting and remain conservative, I estimate that Smart Grid–enabled EE beyond the other adjustment categories is 4% of 2030 sales, or 181 billion kWh. This estimate is about 80% of EPRI's range and far below ACEEE's estimate. As above, to the extent this is overly conservative the prospects for sales growth diminish, and vice-versa.

Onsite Distributed Generation

Distributed generation (DG) made on the customer's side of the meter is certainly part of electricity use, but it is not electricity sold by utilities unless the utility actually owns the onsite generator. Although utilities ultimately may own a substantial portion of onsite DG (see Chapter 13), for the purposes of this calculation it is easiest to assume that all onsite power is not counted as utilities' sales.[13]

There are many policy changes that may cause distributed power to increase faster than forecasted in the EIA reference case. The current W-M legislation and other federal RES proposals give three renewable energy credits (RECs) for every MWh generated. Nevertheless, EIA's simulation of W-M finds that the legislation will cause an increase of only 3 to 5 billion kWh of onsite DG, increasing it only about 2.5 to 4.0% by 2030.[14] EIA's simulation of the bill shows about 50 billion kWh additional renewables (large as well as small scale) from

W-M but *no* additional cogeneration or gas-fired DG. However, EIA also acknowledges great uncertainty around virtually all of the key dimensions of this analysis—the total need for new power plants (i.e., sales growth), the ability to site and get transmission to large-scale renewables, cost trends for generation, and other variables.

Unfortunately, most other policy simulations do not break out distributed renewables from upstream, large-scale installations. Although photovoltaic (PV) grows substantially in many of these simulations, there is no way to tell how much is located "behind the meter." However, one highly specialized EIA analysis conducted in 2000 was focused entirely on the impact of various policies then under discussion on DG. This analysis was based on DG cost estimates that are now much lower than EIA's current DG costs, as well as a higher projected sales base, so they probably overstate policy impacts. However, they give an indication of the magnitude of the potential impacts of policies that promote DG.

The EIA study examines three scenarios: advanced technologies with lower costs, a nationwide net metering policy for all DG, and a permanent 40% tax credit. The differences from the 2000 AEO reference case for these policies are shown in Table A-5.

These estimates seem too high in view of current information, but they indicate that DG policies can have substantial impacts. As a somewhat nonconservative course, I choose 30 billion kWh as potential 2030 nonutility DG, approximately double the RES impacts modeled by EIA. This assumption reflects a

Table A-5. EIA Study of Distributed Generation Policy Impacts Based on [Obsolete] Year 2000 Base Case (billion kWh)

Policy Scenario	Change in 2020 Utility Sales	Extrapolated Change in 2030 Utility Sales
Advanced Technology	5	9
Net Metering and Advanced Technology	18	33
Advanced Technology and 40% Tax Credit	79	126

Note: 2030 impacts are scaled linearly from 2010 and 2020 results.

Source: Boedecker, Erin, John Cymbalsky, and Stephen Wade, "Modeling Distributed Generation in the NEMS Buildings Models," Energy Information Administration, July 30, 2002. Tables 3–10. Available at: http://www.eia.doe.gov/oiaf/analysispaper/electricity_generation.html.

possible future with strong local encouragement for community-based renewable development and improved availability of low-cost downstream storage beyond that assumed by EIA.

I think these are highly possible outcomes. Sources in California, often a bellwether for energy developments, report a rapid rise in interest by real estate developers installing DG as part of large-scale developments. As the cost differences between DG at the 1 to 2 MW (not the individual household) and 100 MW scale decline, many new large-scale developments will routinely include DG, with increasing penetration especially likely in the latter portion of the forecast.

Total Adjustments to Sales

The adjustments discussed above are positive for PHEVs and negative (i.e., sales-reducing) for the remaining factors. The net impact of all these adjustments is 674 billion kWh. This compares to the total EIA-estimated reference case sales growth of 806 billion kWh between 2008 and 2030.

APPENDIX B

Part 1. Summary of Selected Large-Scale Power Generating Technologies (2008 constant dollars)

		Typical Plant Size Range (MW)	Average Investment Costs 2010	Average Investment Costs 2030	Charge Rate (%)	Capacity Factor (%)	Fuel Cost ($/MWh)	O&M Costs ($/MWh)	Average Cost of Power Excluding Carbon Emissions 2010	2030
		[A]	[B]	[C]	[D]	[E]	[F]	[G]	[H]	[I]
Coal										
Coal—Without CCS	[1]	400	$2,500	N/A	8.28%	85%	$18.32	$6.33	$52	N/A
Coal—With CCS	[2]	600	$4,000	$3,000	8.28%	85%	$21.01	$12.00	$77	$66
Nuclear										
Advanced Nuclear	[3]	1,350	$4,000–$8,000	$4,000	8.02%	91%	$7.19	$9.01	$57–$97	$57
Natural Gas										
Natural Gas—Base ($4/MMBtu Gas)	[4]	600	$1,000	$900	8.28%	85%	$28.94	$2.91	$43	$42
Natural Gas—Base ($6/MMBtu Gas)		600	$1,000	$900	8.28%	85%	$40.51	$2.91	$55	$53
Natural Gas—Base ($10/MMBtu Gas)		600	$1,000	$900	8.28%	85%	$63.66	$2.91	$78	$77
Natural Gas—Peaking Plant ($4/MMBtu Gas)	[5]	230	$750	$650	8.02%	11%	$37.16	$9.12	$107	$99

Part 1. Continued

	Typical Plant Size Range (MW)	Average Investment Costs 2010	Average Investment Costs 2030	Charge Rate (%)	Capacity Factor (%)	Fuel Cost ($/MWh)	O&M Costs ($/MWh)	Average Cost of Power Excluding Carbon Emissions	
								2010	2030
	[A]	[B]	[C]	[D]	[E]	[F]	[G]	[H]	[I]
Natural Gas—Peaking Plant ($6/MMBtu Gas)	230	$750	$650	8.02%	11%	$55.73	$9.12	$125	$117
Natural Gas—Peaking Plant ($10/MMBtu Gas)	230	$750	$650	8.02%	11%	$92.89	$9.12	$162	$154

Renewables: A 30% Investment Tax Credit is accounted for in 2010 average construction costs of Solar power. A $21/MWh Production Tax Credit is ac-counted for in 2010 average construction costs of Wind, Biomass, and Geothermal.

Costs of controllable capacity for backup power and grid integration not included. These costs add roughly 30% to 50%.

Incremental transmission Costs not included.

Solar									
Large Solar PV	[6]	50–260	$3,437	$3,332	7.56 / 7.03%	21%	$0.00	$6.27	$146	$132
Solar Thermal (no storage)	[7]	100	$3,000–$5,800	$3,000	7.56 / 7.03%	21%	$0.00	$30.46	$152–$266	$144
Wind									
Large Wind Farms— Onshore	[8]	50–1,000	$1,576	$1,827	7.03%	35%	$0.00	$9.79	$46	$51
Large Wind Farms— Offshore	[9]	100–1,000	$2,997	$2,983	7.03%	43%	$0.00	$23.81	$80	$80
Other Renewables									
Geothermal	[10]	50	$3,000	$4,143	7.03%	80%	$0.00	$24.44	$54	$66
Biomass Power	[11]	80	$2,899	$2,835	7.22%	77%	$13.78	$11.66	$56	$56

Notes:

[1] and [2]: CCS refers to carbon capture and sequestration.

[4] and [5]: Natural gas costs at $6.00/MMBtu.

[H] and [I]: Calculated as ([B] × [D] × 1000) / ([E] × 8760) + [F] + [G], and ([C] × [D] × 1000) / ([E] × 8760) + [F] + [G], respectively.

CO_2 cost is calculated as [J] × 50 ($ per metric ton).

[1]: [A]: AEO 2009 Assumptions, Table 8.2.

[B]: Author's judgment based upon twenty external studies on coal plant costs without CCS in or around 2010 (such as DOE; AEO 2009 Assumptions; ReEDS; EPRI TAG; and an average of thirty-two new coal projects reported by Ventyx).

[C]: Coal without CCS assumed to be obsolete in 2030.

[D]: Calculated using a twenty-year MACRS tax depreciation schedule. See http://www.irs.gov/irs-pdf/p946.pdf.

[E]: Author estimate.

[F]: Calculated using external studies listed in [B] and [C]. When fuel costs not reported, cost of coal assumed to be $2.00/MMBtu. Heat rate of 9,200 Btu /kWh based on AEO 2009 Assumptions.

[G]: Calculated using external studies listed in [B] and [C]. When O&M costs not reported, AEO 2009 Assumptions are used.

[2]: [A]: AEO 2009 Assumptions, Table 8.2.

[B]: Author's judgment based upon seven external studies on coal plant costs with CCS in or around 2010 (such as AEO 2009 Assumptions; ReEDS; CRS; and an average of nine new coal projects reported by Ventyx).

[C]: Author's judgment based upon nine external studies on coal plant costs with CCS in or around 2030 (such as AEO 2009 Assumptions; ReEDS; EPRI TAG).

[D]: Calculated using a twenty-year MACRS tax depreciation schedule. See http://www.irs.gov/pub/irs-pdf/p946.pdf.

[E]: Author estimate.

[F]: Calculated using external studies listed in [B] and[C]. When fuel costs not reported, cost of coal assumed to be $2.00/MMBtu. Heat rate of 10,781 Btu / kWh based on AEO 2009 Assumptions.

[G]: Calculated using external studies listed in [B] and [C]. When O&M costs not reported, AEO 2009 Assumptions are used. For coal with CCS, O&M includes costs of CO_2 transport and storage.

[3]: [A]: AEO 2009 Assumptions, Table 8.2.

[B]: Cost range is based upon sixteen external studies on nuclear plant costs in or around 2010 (such as DOE; AEO 2009 Assumptions; EPRI TAG; MIT; CRS; NEI; and an average of thirty-four new nuclear projects reported by Ventyx).

Part 1. Continued

[C]: Author's judgment based upon seven external studies on nuclear plant costs in or around 2030 (such as DOE; AEO 2009 Assumptions; and ReEDS).

[D]: Calculated using a fifteen-year MACRS tax depreciation schedule. See http://www.irs.gov/pub/irs-pdf/p946.pdf.

[E]: Calculated from AEO 2009 Assumptions.

[F]: Calculated using external studies listed in [B] and [C]. When fuel costs not reported, cost of uranium assumed to be $0.55/MMBtu. Heat rate of 10,434 Btu/kWh based on AEO 2009 Assumptions.

[G]: Calculated using external studies listed in [B] and [C]. When O&M costs not reported, AEO 2009 Assumptions are used. In addition, O&M includes cost of nuclear spent fuel and decommissioning (approximately $2/MWh).

[4]: [A]: Average of sixty-three new combined-cycle natural gas projects as reported in the June 2009 version of The Velocity Suite from Ventyx, Inc.

[B]: Author's judgment based upon sixteen external studies on natural gas combined cycle plant costs in or around 2010 (such as DOE; AEO 2009 Assumptions; MIT; EPRI TAG; NEI; CRS; and an average of sixty-three new combined cycle gas projects reported by Ventyx).

[C]: Author's judgment based upon eleven external studies on natural gas combined cycle plant costs in or around 2030 (such as DOE; AEO 2009 Assumptions; and ReEDS).

[D]: Calculated using a twenty-year MACRS tax depreciation schedule. See http://www.irs.gov/pub/irs-pdf/p946.pdf.

[E]: Capacity factor of natural gas CC large-scale plant from AEO 2009 Assumptions.

[F]: Calculated using three scenarios of natural gas price. Heat rate of 6,752 Btu/kWh based on AEO 2009 Assumptions.

[G]: Calculated using external studies listed in [B] and [C]. When O&M costs not reported, AEO 2009 Assumptions are used.

[5]: [A]: AEO 2009 Assumptions, Table 8.2 (for advanced combustion turbine).

[B]: Author's judgment based upon nine external studies on natural gas combustion turbine plant costs in or around 2010 (such as AEO 2009 Assumptions; ReEDS; and an average of 40 new combustion turbine gas projects reported by Ventyx).

[C]: Author's judgment based upon ten external studies on natural combustion turbine plant costs in or around 2030 (such as AEO 2009 Assumptions; and ReEDS).

[D]: Calculated using a fifteen-year MACRS tax depreciation schedule. See http://www.irs.gov/pub/irs-pdf/p946.pdf.

[E]: Calculated from AEO 2009 Assumptions.

[F]: Calculated using three scenarios of natural gas price. Heat rate of 9,289 Btu/kWh based on AEO 2009 Assumptions.

[G]: Calculated using external studies listed in [B] and [C]. When O&M costs not reported, AEO 2009 Assumptions are used.

[6]: [A]: Author's estimate.

[B]: Calculated from five external studies on large photovoltaic capital costs in or around 2010 (RETI; Connecticut IRP; MIT; AEO 2009 Assumptions; and an average of three new large photovoltaic projects reported by Ventyx). (See "Sources" for more detail.)

[C]: Calculated from two external studies on large photovoltaic capital costs in or around 2030 (RETI; AEO 2009 Assumptions). (See "Sources" for more detail)

[D]: Calculated using a five-year MACRS tax depreciation schedule. See http://www.dsireusa.org/incentives/incentive.cfm?Incentive_Code=US06F&re=1&ee=0. Tax depreciation basis is reduced by one-half the ITC amount in 2010. 7.56% in 2010; 7.03% in 2030. The tax depreciation basis is reduced by 15% (i.e., one-half the ITC amount) in 2010, resulting in a higher charge rate.

[E] and [G]: Calculated from AEO 2009 Assumptions.

[7]: [A]: AEO 2009 Assumption. Table 8.2.

[B]: Cost range is based upon four external studies on large solar thermal capital costs reported by Ventyx.

[E] and [G]: Calculated from AEO 2009 Assumptions.

[C]: Author's judgment based upon two external studies on large solar thermal capital costs in or around 2030 (AEO 2009 Assumptions; EPRI TAG).

[D]: Calculated using a five-year MACRS tax depreciation schedule. See http://www.dsireusa.org/incentives/incentive.cfm?Incentive_Code=US06F&re=1&ee=0. 7.56% in 2010; 7.03% in 2030. The tax depreciation basis is reduced by 15% (i.e., one-half the ITC amount) in 2010, resulting in a higher charge rate.

[E] and [G]: Calculated from AEO 2009 Assumptions.

[8]: [A]: Author's judgment based upon AEO 2009 Assumptions, Table 8.2.

[B]: Calculated from six external studies on large wind capital costs in or around 2010 (AEO 2009 Assumptions; CRS; MIT; and an average of nineteen new large wind projects reported by Ventyx). (See "Sources" for more detail.)

[C]: Calculated from two external studies on large wind capital costs in or around 2030 (AEO 2009 Assumptions; DOE; CRS; EPRI TAG; RETI; and an average of 185 new large wind projects reported by Ventyx). (See "Sources" for more detail.) Note that consumer-level subsidies, which would reduce capital costs for wind farms, are not included in 2030.

[D]: Calculated using a five-year MACRS tax depreciation schedule. See http://www.dsireusa.org/incentives/incentive.cfm?Incentive_Code=US06F&re=1&ee=0.

[E] and [G]: Calculated from AEO 2009 Assumptions.

[9]: [A]: Author's judgment based upon AEO 2009 Assumptions, Table 8.2.

[B]: Calculated from three external studies on large offshore wind capital costs in or around 2010 (AEO 2009 Assumptions; DOE; and an average of seven new large offshore wind projects reported by Ventyx). (See "Sources" for more detail.)

[C]: Calculated from two external studies on large offshore wind capital costs in or around 2030 (AEO 2009 Assumptions; DOE). Note that consumer-level subsidies, which would reduce capital costs for wind farms, are not included in 2030.

[D]: Calculated using a five-year MACRS tax depreciation schedule. See http://www.dsireusa.org/incentives/incentive.cfm?Incentive_Code=US06F&re=1&ee=0.

[E] and [G]: Calculated from AEO 2009 Assumptions. Capacity factor based on 2015 projection.

[10]: [A]: AEO 2009 Assumptions. Table 8.2.

[B]: Author's judgment based upon seven external studies on geothermal capital costs in or around 2010 (such as CRS; RETI; ReEDS; and an average of twenty-one new geothermal projects reported by Ventyx).

[C]: Calculated from five external studies on geothermal capital costs in or around 2030 (such as AEO 2009 Assumptions; and ReEDS). Note that consumer-level subsidies, which would reduce capital costs for geothermal plants, are not included in 2030.

Part 1. Continued

[D]: Calculated using a five-year MACRS tax depreciation schedule. See http://www.dsireusa.org/incentives/incentive.cfm?Incentive_Code=US06F&re=0.

[E] and [G]: Calculated from AEO 2009 Assumptions.

[11]: [A]: AEO 2009 Assumption. Table 8.2.

[B]: Calculated from four external studies on hydroelectric capital costs in or around 2010 (such as ReEDS; and an average of thirty new hydraulic turbine projects reported by Ventyx).

[C]: Calculated from five external studies on hydroelectric capital costs in or around 2030 (such as AEO 2009 Assumptions; and ReEDS).

[D]: Calculated using a twenty-year MACRS tax depreciation schedule. See http://www.irs.gov/pub/irs-pdf/p946.pdf.

[E]: Calculated from AEO 2009 Assumptions.

[G]: Calculated using external studies listed in [B] and [C]. When O&M costs not reported, AEO 2009 assumptions are used.

[12]: [A]: AEO 2009 Assumptions. Table 8.2.

[B]: Calculated from six external studies on biomass capital costs in or around 2010 (such as AEO 2009 Assumptions; EPRI TAG; MIT; RETI; and an average of two new large biomass steam turbine projects reported by Ventyx).

[C]: Calculated from AEO 2009 Assumptions. Note that consumer-level subsidies, which would reduce capital costs for biomass plants, are not included in 2030.

[D]: Calculated using a seven-year MACRS tax depreciation schedule. See http://www.dsireusa.org/incentives/incentive.cfm?Incentive_Code=US06F&re=0.

[E]: Calculated from AEO 2009 Assumptions.

[F]: Calculated using external studies listed in [B] and [C]. When fuel costs not reported, cost of biomass assumed to be $1.00/MMBtu. Heat rate of 9,646 Btu/kWh based on AEO 2009 Assumptions.

[G]: Calculated using external studies listed in [B] and [C]. When O&M costs not reported, AEO 2009 assumptions are used.

Part 2. Small-Scale (Distributed) Generating Sources (Costs do not reflect consumer-level subsidies or rebates of any kind, except where noted) (2008 constant dollars)

		Typical Plant Size Range (kW)	Current Capital Cost (2008$/kW)	Projected Capital Cost 2030 ($/kW)	Charge Rate	Capacity Factor	Fuel Costs, $6.00 per MMBtu, Current ($/kWh) (b)	O&M Costs, Current ($/kWh)	Average Cost of Power Excluding Carbon Emissions, Backup and Integration 2010	2030	CO$_2$ Emissions (lbs. per MWh)
		[A]	[B]	[C]	[D]	[E]	[F]	[G]	[H]	[I]	[J]
Natural Gas Microturbine	[1]	30–250	$3,013	$1,358	7.03%	0.70	$0.04	$0.02	$92	$73	1,584.00
Natural Gas Industrial CHP	[2]	10,000	$1,267	$761	7.03%	0.85	$0.02	$0.01	$39	$34	1,166.00
Large Solar Distributed PV	[3]	10,000–20,000	$3,437	$3,332	7.03%	0.21	$0.00	$0.01	$136	$132	0.00
Small Solar Distributed ("Rooftop") PV	[4]	4.5	$8,828	$8,559	7.03%	0.21	$0.00	$0.01	$339	$329	0.00
Large Wind Farms—Onshore	[5]		$1,576	$1,827	7.03%	0.35	$0.00	$0.01	$46	$51	0.00
Residential Distributed Wind	[6]		$4,794	$3,174	7.03%	0.20	$0.00	$0.01	$202	$137	0.00

Part 2. Continued

	Typical Plant Size Range (kW)	Current Capital Cost (2008$/kW)	Projected Capital Cost 2030 ($/kW)	Charge Rate	Capacity Factor (b)	Fuel Costs, Current ($/kWh)	O&M Costs, Current ($/kWh)	Average Cost of Power Excluding Carbon Emissions, Backup and Integration 2010	2030	CO₂ Emissions (lbs. per MWh)
						$6.00 per MMBtu				
	[A]	[B]	[C]	[D]	[E]	[F]	[G]	[H]	[I]	[J]
Fuel Cells—noncogeneration [7]	2,000	$7,980	$3,988	7.03%	0.9	$0.05	$0.05	$173	$137	0.02–0.06
Fuel Cells—noncogeneration [8]	10	$10,041	$3,988	7.03%	0.9	$0.05	$0.05	$191	$137	0.02–0.06

Notes:

[D]: All technologies use a charge rate of 7.03% copied from the charge rate for large PV.

[F]: These fuel costs include a credit for heat, lowering them by 50%, for the two cogeneration technologies.

[H] and [I]: Calculated as ([B] × [D] × 1000) / ([E] × 8760) + [F] × 1000 + [G] × 1000, and (([C] × [D] × 1000) / ([E] × 8760) + [F] × 1000 + [G] × 1000, respectively.

CO₂ cost is calculated as [J] × 50 ($ per metric ton) / 2,200 (lb. per metric ton).

[I]: [B]: Average of 30 and 65 kW capacity costs per kW from EPA source shown, increased from 2007 to 2008 prices by the Producer Price Index (PPI) provided by the U.S. Department of Labor: Bureau of Labor Statistics.

[C]: Scaled to 2030 using ratio of $1.05/kW 2030 to $2.33/kW 2010 from http://www.nrel.gov/docs/fy04osti/34783.pdf as below.

[E]: Author assumption.

[F]: Uses electric conversion efficiency of 25% from http://www.energy.ca.gov/distgen/equipment/microturbines/performance.html. NOTE FUEL COSTS REDUCED BY 50% TO REPRESENT HEAT CREDIT.

[G]: Midpoint value for microturbine O&M from Table 2 p. 1–8 of http://www.nrel.gov/docs/fy04osti/34783.pdf.

[J]: Source: "Power Plants: Characteristics and Costs," Congressional Research Service. Note that CO₂ emissions rate is not adjusted to account for the fact that useful heat is also provided.

[2]: [C]: 2030 derived by scaling 2008 value by the ratio of 2030 cost per kW to 2010 cost per KW from 2009 AEO assumptions for natural gas engine—$1.13 in 2030 and $1.88 in 2010 (ignoring the gap between 2008 and 2010). Underlying data from http://www.nrel.gov/docs/fy04osti/34783.pdf.

[E]: Capacity factor of natural gas CC large scale plant from AEO 2009 Assumptions.

[F]: Uses a 6,752 heat rate from natural gas base load AEO 2009 Assumptions. NOTE FUEL COSTS REDUCED BY 50% TO REPRESENT HEAT CREDIT.

[G]: Midpoint estimate from gas turbine commercial from NREL Table 2 p. 1–8.

[J]: Source: "Power Plants: Characteristics and Costs," Congressional Research Service... Note that CO₂ emissions rate is not adjusted to account for the fact that useful heat is also provided.

[3]: [B]: Calculated from five external studies on large photovoltaic capital costs in or around 2010 (RETI; Connecticut IRP; MIT; AEO 2009 Assumptions; and an average of three new large photovoltaic projects reported by Ventyx). (See "Sources" for more detail.)

[C]: Calculated from two external studies on large photovoltaic capital costs in or around 2030 (RETI; AEO 2009 Assumptions). (See "Sources" for more detail.)

[E] and [G]: Calculated from AEO 2009 Assumptions.

[4]: [A]: In DC watts.

[B]: Systems 2–5 kW, increased from 2007 to 2008 prices.

[C]: Reduced by the ratio of large PV 2030 constant costs to large PV 2010 costs from row [3].

[G]: Used large-scale PV O&M costs (see row [3]), but true rooftop O&M costs likely to be much higher.

[5]: [B]: Calculated from six external studies on large wind capital costs in or around 2010 (AEO 2009 Assumptions; DOE; CRS; EPRI TAG; RETI; and an average of 185 new large wind projects reported by Ventyx). (See "Sources" for more detail.)

[C]: Calculated from two external studies on large wind capital costs in or around 2030 (AEO 2009 Assumptions; DOE). (See "Sources" for more detail.) Note that consumer-level subsidies, which would reduce capital costs for wind farms, are not included in 2030.

[E] and [G]: Calculated from AEO 2009 Assumptions.

[6]: [B] and [C]: Consistent with an AWEA survey reporting $3 to $6 per watt.

[E]: Author assumption.

[7]: [A]: "Catalog of CHP Technologies," U.S. Environmental Protection Agency; Table 2.

[E]: "Catalog of CHP Technologies," U.S. Environmental Protection Agency; Table 2.

[F]: Midpoint electric efficiency from "Catalog of CHP Technologies," U.S. Environmental Protection Agency; Table 2.

Part 2. Continued

[G]: AEO 2009 Assumption. Table 2 p. 1–8 cite "Catalog of CHP Technologies," U.S. Environmental Protection Agency shows a lower 3 cents/kWh.

[J]: EPA study, "Technology Characterization: Fuel Cells," December 2008. See http://www.epa.gov/CHP/documents/catalog_chptech_fuel_cells.pdf .

[8]: [A]: "Catalog of CHP Technologies," U.S. Environmental Protection Agency, Table 2.

[E]: "Catalog of CHP Technologies," U.S. Environmental Protection Agency, Table 2.

[F]: Midpoint electric efficiency from "Catalog of CHP Technologies," U.S. Environmental Protection Agency, Table 2.

[G]: AEO 2009 Assumption. Table 2 p. 1–8 cite "Catalog of CHP Technologies," U.S. Environmental Protection Agency, shows a lower 3 cents/kWh.

[J]: EPA study, "Technology Characterization: Fuel Cells," December 2008. See http://www.epa.gov/CHP/documents/catalog_chptech_fuel_cells.pdf .

Sources:

AEO 09   "Assumptions to the Annual Energy Outlook With Projections to 2030," Energy Information Administration, March 2009. Available at: http://www.eia.doe.gov/ oiaf/aeo/assumption/index.html.

CEC   Ken Darrow, Bruce Hedman, and Anne Hampson, "CHP Market Assessment," Prepared for the California Energy Commission by ICF International, July 23, 2009.

CRS   "Power Plants: Characteristics and Costs," Congressional Research Service, November 13, 2008. Available at: http://assets.opencrs.com/rpts/RL34746_20081113.pdf.

CT IRP   "Integrated Resource Plan for Connecticut," by Marc Chupka, Ahmad Faruqui, Dean M. Murphy, Samuel A. Newell, and Joseph B. Wharton, *The Brattle Group*, Inc., January 1, 2008.

DOE   "20% Wind Energy by 2030: Increasing Wind Energy's Contribution to U.S. Electricity Supply," Energy and Efficiency and Renewable Energy, July 2008.

EB 08   Colette Lewiner, "The Global Nuclear Renaissance," *Energy Biz*, March/April 2008.

EB 09   Martin Rosenberg, "Bring on Nuclear," *Energy Biz*, March/April 2009.

EEI   Marc Chupka et al., "Transforming America's Power Industry: The Investment Challenge 2010–2030," *The Brattle Group*, November 2008.

EERE   "Renewable Energy Data Book," U.S. Department of Energy, September 2008. Available at: http://www1.eere.energy.gov/maps_data/pdfs/eere_databook_ 091208.pdf.

EPA   "Catalog of CHP Technologies," U.S. Environmental Protection Agency, December 2008.

MIT 06   "Future of Geothermal Energy: Impact of Enhanced Geothermal Systems (EGS) on the United States in the 21st Century," MIT Report, 2006.

MIT 09   "Update of the MIT 2003 Future of Nuclear Power," MIT Energy Initiative, 2009. Available at http://web.mit.edu/nuclearpower/pdf/nuclearpower-update2009.pdf.

NEI   "The Cost of New Generating Capacity," Nuclear Energy Institute, February 2009.

ReEDS   "ReEDS Base Case Data," National Renewable Energy Laboratory, 2008. Available at http://www.nrel.gov/analysis/reeds/pdfs/reeds_chap_2.pdf.

RETI   "Renewable Energy Transmission Initiative Phase 1 B—Final Report," Black and Veatch, January 2009. Available at: http://www.energy.ca.gov/reti/documents/index.html.

Severance   Craig A. Severance, "Business Risks and Costs of New Nuclear Power," posted on the Climate Progress Blog, 2009. Available at http://climateprogress.org/wp-content/uploads/2009/01/nuclear-costs-2009.pdf.

TAG   "Program on Technology Innovation: Integrated Generation Technology Options," The Electric Power Research Institute, November 2008. Available at: http://mydocs.epri.com/docs/public/000000000101018329.pdf.

Ventyx   The costs represented come from a variety of sources, including, but not limited to, press releases, news articles, permits, Web sites, and company contacts—as reported in the June 2009 version of *The Velocity Suite* from Ventyx, Inc.

APPENDIX C

Further Discussion and Reading Regarding Competition in the Power Industry

TWO PRIMARY explanations have been advanced to explain the apparent gap between deregulation's rhetoric and reality in the power industry. The first is that competition was never a very sound idea in this industry, for equity as well as for economic reasons. This view was espoused by economists such as Harry Trebing and regulatory scholars like Judge Richard Cudahy, as well as by parts of the industry itself.

More recent papers that are either skeptical of the case for deregulation or entirely unconvinced come from Professors Borenstein, Bushnell, and Brennan, as well as Richard Rosen, Jacqueline Lang Weaver, and John Kelly. My own writings have also often been cautionary, noting that the case for retail choice (as distinct from wholesale competition) was weak and that competitive conditions in many markets were questionable.

The second explanation is described by policy analysts like Adam B. Summers of the Reason Foundation as "a government regulation failure, not a market

failure." This argument holds that politicians have not been able to stop themselves from placing controls on what were supposed to be deregulated markets
that prevent them from functioning properly. Articulations of this view are
found in recent books by James Sweeney and Charles Cicchetti and several articles by Professor Paul Joskow.

As the following studies show, there is a degree of truth in both schools of
thought. There were deep and pervasive flaws in the way we implemented electric competition in this country, wholesale as well as retail. The implementation
flaws were exacerbated by inadequate legislative and regulatory responses. Political leaders found it impossible to allow electric prices to rise rapidly and unpredictably, as they do in other deregulated markets. At the same time, electricity is
inherently ill-suited to be completely deregulated, and, where competition is appropriate, the markets work only with strong, highly skilled monitoring and
enforcement.

Partial Bibliography: Studies of Wholesale and Retail Competition

A cross section of these studies includes the following scholarly works and policy papers.

Apt, Jay. "Competition Has Not Lowered U.S. Industrial Electricity Prices," *Electricity
Journal*, March 2005: 52–56.

Barmack, Matthew, Edward Kahn, and Susan Tierney. "A Cost–Benefit Assessment of
Wholesale Electricity Restructuring and Competition in New England," Analysis
Group, 2006.

Behr, Peter. "Probe of California Energy Crisis Facing Hurdles," *Washington Post*, January
11, 2003: E01.

Blumsack, Seth A., Jay Apt, and Lester B. Lave. "Lessons from the Failure of U.S. Electricity Restructuring," *Electricity Journal*, March 2006: 15–32.

Blumstein, Carl, L. S. Friedman, and R. J. Green. "The History of Electricity Restructuring in California." CSEM WP 103, UCEI, August 2002.

Borenstein, Severin, and James Bushnell. "Electricity Restructuring: Deregulation or
Reregulation?" *Regulation*, 23 (2), 2000.

Brennan, Tim. "Questioning the Conventional 'Wisdom,'" *Regulation*, 24, (3), Fall 2001.

Brown, Matthew, H. "California's Power Crisis: What Happened? What Can We Learn?"
National Conference of State Legislatures, March 2001.

Bushnell, James B., and Frank A. Wolak. "Regulation and the Leverage of Local Market Power in the California Electricity Market," July 1999.

"California Energy Crisis: Causes and Solutions," Union of Concerned Scientists, http://www.ucsusa.org/clean_energy/renewable_energy/page.cfm?pageID=68.

Chandley, John D., Carl R. Danner, Christopher E. Groves, et al. "California's Electricity Markets: Structure, Crisis, and Needed Reforms," LECG, LLC., January 17, 2003.

Cicchetti, Charles J., and Colin M. Long, with Kristina M Sepetys. *Restructuring Electricity Markets: A World Perspective Post California and Enron.* Visions Communications, 2003.

Cudahy, Richard D. "The Coming Demise of Deregulation." *Yale Journal on Regulation*, Issue 10.1, 1993.

Faruqui, Ahmad, Hung-po Chao, Vic Niemeyer, Jeremy Platt, and Karl Stahlkopf. "Analyzing California's Power Crisis," *Energy Journal*, 22 (4), 2001: 29–51.

Harvey, Scott M., Bruce M. McConihe, and Susan L. Pope. "Analysis of the Impact of Coordinated Electricity Markets on Consumer Electricity Charges." Draft, LECG, November 20, 2006.

Fox-Penner, Peter. "A Welcome Truce in the Electricity Wars." *Public Utilities Fortnightly*, 48, 2005: 51.

Fox-Penner, Peter. "Electric Power Deregulation: Blessings and Blemishes: A Non-Technical Review of the Issues Associated with Competition in Today's Electric Power Industry." Prepared for the National Council on Competition and the Electric Industry, March 14, 2000.

Fox-Penner, Peter. *Electricity Utility Restructuring.* (Vienna, Virginia: Public Utilities Reports), 1997.

Hogan, William W. "WEPEX: Building the Structure for a Competitive Electricity Market." Presented at the FERC Technical Conference Concerning WEPEX, Washington, DC, August 1, 1996.

Joskow, Paul L. "California's Electricity Crisis." *Oxford Review of Economic Policy*, 2001: 365–388.

Joskow, Paul L. "Regulatory Failure, Regulatory Reform and Structural Change in the Electric Power Industry," *Brookings Papers on Microeconomic Activity*, Special Issue, 1989: 125–220, 1989.

Kahn, Michael, and Loretta Lynch. "California's Electricity Options and Challenges Report to Governor Gray Davis," California Public Utilities Commission, August 2, 2000.

Kelly, John. "EMRI: Do Competition and Electricity Mix?" American Public Power Association, 2007.

Kelly, Susan, and Diane Moody. "Wholesale Electric Restructuring: Was 2004 the 'Tipping Point'?" *Electricity Journal*, March 2005: 11–18.

Kleit, Andrew N., and Dek Terrell. "Measuring Potential Efficiency Gains from Deregulation of Electricity Generation: A Bayesian Approach," *Review of Economics and Statistics*, August 2001: 523–530.

Kwoka, John E. "Restructuring the U.S. Electric Power Sector: A Review of Recent Studies." Prepared for American Public Power Association, November 2006.

Mansur, Erin T. "Measuring Welfare in Restructured Electricity Markets," *Review of Economics and Statistics*, May 2008: 369–386.

Markiewicz, Kira, Nancy Rose, and Catherine Wolfram. "Has Restructuring Improved Operating Efficiency at US Electricity Generating Plants?" CSEM WP 135, UCEI, July 2004.

McCullough, Robert. "California Electricity Price Spikes: Factual Evidence," January 15, 2003. http://www.mresearch.com/pdfs/76.pdf.

McCullough, Robert. Memorandum to McCullough Research Clients, "C66 and the Artificial Congestion of California Transmission in January 2001," November 29, 2002.

Moody, Diane. "The Use—and Misuse—of Statistics in Evaluating the Benefits of Restructured Electricity Markets," *Electricity Journal*, March 2007: 57–62.

Morrison, Jay A. "The Clash of Industry Visions," *Electricity Journal*, January/February 2005: 14–30.

Pope, Susan L. "California Electricity Price Spikes: An Update on the Facts," Harvard Electricity Policy Group, December 9, 2002.

"Report to Congress on Competition in the Wholesale and Retail Markets for Electric Energy." Draft, The Electric Energy Market Competition Task Force and FERC, June 5, 2006.

Rosen, Richard. "Regulating Power: An Idea Whose Time Is Back." *American Prospect*, 2002: 22.

Summers, Adam B. "Call for Real Electricity Deregulation." *Ventura County Star*, 2005. Available through The Reason Foundation Web site: http://reason.org/news/show/call-for-real-electricity-dere.

Sweeney, James L. *The California Electricity Crisis* Hoover Institution Press, 2002: Chap. 2, 4.

Taber, John, Duane Chapman, and Tim Mount. "Examining the Effects of Deregulation on Retail Electricity Prices." WP2005-14, February 2006.

"The Competitive Retail Electricity Markets and Customer Protections: Are Customers Really Better Off?" Regulatory Compliance Services, March 2009.

Thornberg, Christopher F. "Of Megawatts and Men: Understanding the Causes of the California Power Crisis." The UCLA Anderson Forecast, 2002.

Tierney, Susan F. "Decoding Developments in Today's Electric Industry—Ten Points in the Prism." Analysis Group, October 2007.

Trebing, Harry M. "The Networks as Infrastructure—The Reestablishment of Market Power." *Journal of Economics Issues*, 28, 1994.

Weare, Christopher. "The California Electricity Crisis: Causes and Policy Options," Public Policy Institute of California, 2003.

Weaver, Jaqueline Lang. "Can Energy Markets Be Trusted? The Effect of the Rise and Fall of Enron on Energy Markets." *Houston Business and Tax Law Journal*, 2004.

Wolak, Frank A. "What Went Wrong with California's Re-structured Electricity Market? (And How to Fix It)." Presented at the Stanford Institute for Economic Policy Research, February 15, 2001.

Notes

The First Electric Revolution

1. Nye, *Electrifying America: Social Meanings*, 1995: 13–14.
2. Goodell, *Big Coal*, 2007: 101.
3. Nye, *Electrifying America: Social Meanings*, 1995: 236.
4. Insull, *Central Station Electric Service*, 1915: 119–120.
5. EIA, 2009(a); and "TVA Goes to War," 2009.
6. Clark, *Energy for Survival*, 1974.
7. Gartner, Inc., "Gartner Says," 2009.
8. EIA, 2009(b).
9. DOE, "The Smart Grid: An Introduction."
10. National Academy of Engineering, "Greatest Engineering."
11. Pachauri and Reisinger, (eds.) "Climate Change 2007: Synthesis Report," 2007.
12. "Economic Security: Will the Third Oil Shock Be the Charm?" Clean Fuels Development Coalition, September 2009.
13. Friedman. "Statement of the Union of Concerned Scientists before the House Committee on Energy and Commerce Subcommittee on Energy and Environment," April 24, 2009.
14. Gronewold, "IPCC Chief Raps G-8," 2009.
15. Fox-Penner, "Transforming America's Power Industry," 2008.
16. Jonnes, "Empires of Light," 2003.

Deregulation, Past and Prologue

1. Although it is now hard to imagine, this was an era in which Enron was widely seen as one of the most innovative and successful companies in the world. The readers of *Fortune* voted it America's most respected company for six consecutive years, and its influence on policymakers across the political spectrum was enormous. See Claeys, "Changing Course: Latest RKS," 2007.
2. Dao, "The End of the Last Great Monopoly," 1996.
3. U.S. House Committee on Commerce, 1996.
4. Beder, "Electricity Deregulation Con Game," 2003.
5. "DOE: Residential Consumers," 1999.
6. CAISO, "Alerts, Warnings," 2004, and "CAISO Urges," 2001.
7. Sheffrin, memorandum to Market Issues/ADR Committee, CAISO, 2001(a) and memorandum to CAISO Board of Governor," 2001(b).
8. Kurt, "California Power Crunch Sends," 2001.
9. Pfeifenberger, Basheda, and Schumacher, "Restructuring Revisited." 2007: 68.
10. To clarify, the enormous increases in POLR rates were caused overwhelmingly by increases in the cost of making electricity from fuels, labor, and power equipment. The problem for policymakers, however, was that they had promised that deregulation would lower prices *relative to continued regulation*. Since about half the states deregulated and the other half did not, it was easy to see which category of states had experienced lower rate increases during this period. While many perceived that rates had actually increased faster in deregulated states, analysis by my *Brattle* colleagues (note 3 supra) show that the pace of rate increases was essentially identical over the period.

    There are many reasons why rates did not decline more in deregulated states. The deregulated states were the more expensive states to begin with, where the costs of doing anything are higher. Moreover, deregulated markets are *designed* to set the price of power at the cost of the most expensive power plant operating in any hour, whereas regulated prices are set at the average of high- and low-price plants operating in that same hour. (In the next chapter, we shall see that the Smart Grid is removing this difference between regulated and deregulated rates, but the political damage to deregulation's reputation has long since occurred.) Market power and poor market design also contributed to higher prices in some deregulated markets, although much of this has now been removed.

    In Chapter 11 we return to the question of competition's role in the future structure of the industry. For more discussion and analysis of the impact of competition on the industry, see Appendix C.
11. At this point, regulators and consumer representatives were unbridled in their outrage against deregulation. "The forced experiment called 'deregulation' was a tri-

umph of ideology over the needs of average citizens," said New York legislator Richard Brodsky in 2007, as he introduced legislation to reregulate his state's power markets. Full page ads published by the Consumer Coalition asserted that, "The promised benefits of electricity deregulation have failed to materialize." Charles Acquard, executive director of the National Association of State Utility Consumer Advocates observed, "Nobody's benefited from deregulation—period, end of story."

12. Davis, Paletta, and Smith, "Unraveling Reagan: Amid Turmoil, U.S. Turns Away from Decades of Deregulation," 2008. For a contrasting view see Winston, "Day of Reckoning for Microeconomists," Sept., 1993: 1263–1289.

13. I argued this point in 2008 in testimony before the Colorado Public Utilities Commission, Docket No. 08S-520E. MIT technology policy experts Charles Weiss and William Bonvillian make a similar observation in the context of technology policy recommendations: "Technologies like biofuels and carbon capture and sequestration are also secondary technologies, components to established energy platforms or systems. However, they face immediate political and nonmarket economic competition from established firms that are not likely to accept them. They will require attention to all stages of the innovative process, from research to development to prototyping to demonstration to incentives for market entry. Their implementation will be unlikely without some form of government regulation or mandate."

The New Paradigm

1. The project is described in more detail in Hammerstrom et al., "Pacific Northwest GridWiseTM Testbed 2007" and Chassin and Kiesling, "Decentralized Coordination," 2008. Historical information on Gridwise is available at "How Did Gridwise Start?" Gridwise and PNNL, gridwise.pnl.gove/foundations/history.stm. For representative press coverage, see Carey, "A Smarter Electric Grid," 2008, and Lohr, "Digital Tools Help Users," 2008. I am indebted to Steve Hauser, Rob Pratt, and Jerry Brous for oral information on the experiment.

2. Appendix 2A of my 1997 book, *Electric Restructuring*, contains a good summary of the economic benefits of interconnection.

3. Rowland, "Get Smart," 2009.

4. Amin, "For the Good of the Grid," 2008: 48–59.

5. Interestingly, a company known as Better Place is promoting essentially this model for electric cars. For the rest of our power uses, however, we are many decades away from storage that has the capacity we need to run our homes in a small enough package and at an affordable price.

Smart Electric Pricing

1. Based on Karl Stahlkopf's (Hawaiian Electric Power) comments as a panelist on "Emerging Smart Grid Technologies: The Future of U.S. Power Distribution," moderated by Peter Fox-Penner, Webinar presented to 2degrees Intelligent Grids Network April 22, 2009, and Ahmad Faruqui, "Creating Value Through Demand Response," *The Brattle Group*, presented at Smart Grid E-Forum, October 23, 2008. Other demand response programs that do not involve hourly prices—though they always include some form of incentive payment—include interruptible or curtailable rates, demand bidding or buyback programs, and emergency demand response. For example, the Hawaii Electric Company has a system that controls 250,000 electric hot water heaters. When power supplies are short on the island, they can turn off some or all of the heaters for brief periods to regain system balance.

2. This concept of critical peak pricing, the "stick," can also be turned around and made into a "carrot" in the form of a rebate when a customer reduces use below some defined, typical levels during the same advance-notice periods.

3. Faruqui, "Creating Value through Demand Response," 2008.

4. "Whirlpool Commits to Smart Grid–Compatible Appliances," 2009.

5. "Baltimore Gas and Electric Company," 2009.

6. "Quantifying Demand Response Benefits in PJM," 2007. The ratio of energy savings between shifters and nonshifters depends on the size of the market and the number of customers who are paying spot prices one way or another through their power bill. This example is particularly large because PJM is the world's largest spot market, and many customers have prices indexed to PJM's hourly prices (known as locational marginal prices, or LMPs). Nonetheless, energy savings have been found to be significant in many different types of market structures, including regulated systems, for both shifting and nonshifting customers, and as noted in the text capacity savings, which also eventually flow to both shifting and nonshifting customers, may be larger. For an extensive review of the results of dynamic pricing, see Faruqui and others, "Moving Toward Utility-Scale," 2009; Cappers, Goldman, and Kathan, "Demand Response in U.S. Electricity," 2009; Smart Grid: The Value Proposition for Consumers E-forum presentations, U.S. Department of Energy, 2008(a), Hosted by National Electric Manufacturers Association, October 23, 2008; Faruqui, "Creating Value through Demand Response," 2008; and Faruqui, Hledik, and Tsoukalis, "The Power of Dynamic Pricing," 2009: 42–56.

7. See Figure 3, Faruqui et al., "Moving toward Utility-Scale," 2009.

8. NERC, "2008 Long-Term Reliability," 2008.

9. It is harder to do this in deregulated markets, as capacity prices (i.e., the price of the 12-cent plant in the example) are set by a market for the short term and are quite

volatile. In regulated markets, the cost of the next power plant doesn't change much, and the calculation of peak price is easier to make with confidence.

10. Faruqui, "Inclining toward Efficiency," 2008: 22–27.

11. Faruqui et al., "Moving toward Utility-Scale," 2009. Section 6 of this report has an excellent summary of the benefits of AMI and how utilities convey the benefits of AMI to regulators—the "business case for AMI." Another good reference in this area is "Deciding on 'Smart' Meters: The Technology Implications of Section 1252 of the Energy Policy Act of 2005," prepared by Plexus Research, Inc., for Edison Electric Institute, September 2006.

12. Andy Satchwell, a consumer advocate working for the state of Indiana, listed six concerns regarding utility Smart Grid investments: Six significant barriers: *resource allocation by utility*: complicated process requiring coordination and technical expertise; *cost of Smart Grid systems*: cost increases dramatically based on communication technology and software system integration; *rapidly changing technologies*: what is a prudent investment today?; *stranded assets*: changes in technology and uncertainty of customer response; *regulatory risks*: cost recovery and unknown payback periods; and *tracking costs and benefits*: complex and lack standard measurements; and see Faruqui, "Breaking Out of the Bubble," 2007: 46–48, 50–51.

13. It is true that many Smart Grid advocates emphasize the essential value of hourly pricing in much the same way the advocates of retail competition once did the same for deregulated pricing. They contrast a Smart Grid in which prices perform a self-balancing function as inherently far more efficient than a regulated system run without prices. (See, for example, Kiesling, "Project Energy Code—Markets," 2009). Much of this volume constitutes my explanation of why I think it is incorrect to align retail deregulation and dynamic pricing. While I see enormous value in dynamic pricing, I do not think retail deregulation will increase because of it. If anything, my view is that state regulation must play a central role in price-setting in order to "get the prices right"—the converse of the more common economic view of regulation. In this case, the proximate technical reason is that unregulated markets will not include the price of generation, transmission, or distribution capacity. These are system savings that only systemic processes can estimate and include—or, in more formal language, these are quasi-public goods. Regulators should do a good job of estimating them, but leaving them out of the price signal is clearly worse than estimating and including them.

14. Greenberg, "The Smart Grid vs. Grandma," 2009. Further opposition to large-scale investment in smart meters is described in a recent *Wall Street Journal*, Smith, "Smart Meter, Dumb Idea?" 2009.

15. In efficiency terms, there is no question: there is nothing economically rational about flat rates. Customers who use power during off-peak periods, but who pay a flat rate that blends cheaper off-peak power and pricey on-peak power to make the

flat rate average are subsidizing customers who use more than their share of power during the on-peak periods. Remember the analogy in the last chapter where everyone paid for their groceries by the pound at the checkout counter regardless of what they put in their cart? If the grocery store is charging a single price per pound for groceries that covers its costs, the people who buy expensive heavy foods (steak) will be subsidized by the people who buy cheap heavy foods (bulk potatoes).

The debate really surrounds the fairness aspect of dynamic prices. Is it fair to the restaurant owner with the lunch trade to pay the same per kilowatt-hour as the owner of the all-night café, even if the costs of serving them are different? There are instances in which society has mandated that prices not reflect different costs for fairness reasons. However, it does seem that there are many more markets in which it isn't perceived as fair when two customers pay the same price for two products that cost different amounts to make, and there does not seem to be any special social attribute of peak-power users that calls for special treatment. Faruqui and Hledik discuss the subject more completely in "Transition to Dynamic Pricing," 2009: 27–33 and Faruqui, "Breaking Out of the Bubble," 2007: 46–48, 50–51.

16. For a very useful overview, see Brockway, "Advanced Metering Infrastructure," 2008 and Faruqui and Wood, "Quantifying the Benefits," 2008.

17. "Smart Meters—Got Bugs?" 2009.

18. Fred Butler, chairman of the National Association of Regulatory Utility Commissioners (NARUC) as well as the New Jersey state regulatory agency, as reported in *McClatchy-Tribune Regional News*, Vock, "Smart Grid's Growth Now," 2009. The complete quote is worth noting:

> You can't have a smart grid and dumb rates. We have been used to—for over 100 years—rates that are the same all day, every day. That's not the way electricity is produced.

19. Universal time-based pricing has been implemented for all large commercial and industrial customers in many states, but mandating that all residential customers pay dynamic prices, with no exceptions, has so far proven to be impossible.

As an alternative, regulators are starting to seriously consider making one form of dynamic pricing (for example, time-of-use rates or peak-time rebate) the default rate but allowing residential customers to opt out of them, with only the submission of the right paperwork or email. Relying on the behavioral principles of choice architecture in the face of the strong status quo effect, described in Thaler and Sunstein's *Nudge* (2008), experience to date suggests that enough customers will remain on the default rates to reap the benefits of dynamic prices, as many as 50%. As explained in the text, it is not essential that every customer experience dynamic prices, only that a sufficient number who really can react so as to shift aggregate demand to a more constant and therefore cheaper time profile of demand.

The Regulatory Mountain

1. "Empowering Consumers," without notes, 2009.
2. For a good, though not extremely recent, review of avoided cost practices, see Beecher, "Avoided Cost: An Essential," 1995. Many state public service commissions have by now developed very sophisticated ways of measuring generation avoided costs. The California Public Utility Commission's (CPUC's) methods are explained in "Methodology and Forecast," 2004, and "Interim Option on E3 Avoided," 2004; and E3, "CEE Cost-Effectiveness Tools," 2007.
3. Centolella and Ott, "The Integration of Price Responsive Demand into PJM Wholesale Power Markets and System Operations," March 9, 2009; 4.
4. In addition to the issues raised by the two PJM experts referenced above, a good example of the ongoing debate occurred again in PJM when sellers complained that market rules overpaid those who used DR to reduce peak demand. See Hogan, "Providing Incentives for Efficient Demand Response," October 29, 2009.
5. *Small Is Profitable* by Lovins et al., discusses locational IRP in Section 1.4.
6. Interstate Renewable Energy Council, 2009.
7. Lovins, et al. *Small Is Profitable*. For the sake of full disclosure, I reviewed this book and my quote is on the original jacket cover. More recent cost information on DG technologies, including recent data from RMI, is reviewed in Chapter 9.
8. For an overview, see http://standards.ieee.org/announcements/bkgnd_stdsprocess .html.
9. The initial roadmap document is EPRI's "Interim Smart Grid Roadmap," 2009. For two other useful overviews of Smart Grid standards issues, see "Introduction to Interoperability and Decision-Maker's," 2007; Gunther and others, "Smart Grid Standards Assessment," 2009; and "Smart Grid Standards Adoption," 2009.
10. Condon, "Lack of Standards," 2009, and see Testimony of Gallagher, before the Committee on Energy and Natural Resources, U.S. Senate, 2009. Also see Katie Fehrenbacher's May 17 post on earth2tech.com, "How to Hammer Out," 2009, and Highfill and Shah. "A Voice for Smart-Grid," 2009.
11. http://www.waterefficiency.net/the-latest/aclara-wifi-standards.aspx.
12. Boyle, "A Smarter, Greener Grid," 2008.
13. Brown, Testimony before the Subcommittee on Energy, 2009.
14. Personal communication, July 23, 2009.
15. "Smart Meters—Got Bugs?" 2009; Goodin, "Buggy 'Smart Meters'," 2009; and Seltzer, "Drive-By-Blackouting," 2009.
16. "Energy Security Wars: Grids vs. Hackers," 2009.
17. Gorman, "Electricity Grid in U.S. Penetrated by Spies," 2008.
18. In *The Future of the Internet*, law professor Jonathan Zittrain argues that security on the Web will ideally occupy a middle ground between a completely uncontrolled environment where, in the words of reviewer David Reed, "robotic spies and

saboteurs [are] everywhere you click" or ubiquitous governmental enforcement and regulation that will squelch creativity. In grid security, however, the balance is likely to shift heavily toward much stronger security provisions and enforcement. See Reed, "Fouling Our Own Net," 2009.

The (Highly Uncertain) Future of Sales

1. U.S. Census Bureau, 2008.
2. EIA, 2009(c).
3. Based on 10–15 million in sales of cars and trucks in 2007; a popular estimate among various sources. See Morgan & Company, Inc., 2008.
4. Figure 4-4 from EPRI's report, "Environmental Assessment of Plug-In Hybrid Electric Vehicles, Volume 1: Nationwide Greenhouse Gas Emissions," uses 15 million vehicles/year. See EPRI, 2007(a).

    Morgan & Company have recorded 2007 sales at roughly 15 million. Richard Cooper, vice president at J.D. Power, has recently estimated auto sales will reach 15 million by 2012–2014, and economists such as Diane Swonk and Mark Zandi have made similar projections. Google's "Clean Energy 2030," report also uses this number as a benchmark for calculating future PHEV penetration.
5. See EPRI, 2007(a).
6. Letendre, Denholm, and Lilienthal, "Plug-In Hybrid and All-Electric Vehicles: New Load or New Resource?" 2006.
7. EIA, 2009(d).
8. Based on net internal U.S. peak demand for 2008 and 2017. See NERC, 2008.
9. EPA, "EPA Analysis," 2009.
10. EIA, 2009(d).
11. EPRI, 2009(a).
12. Schlegel, "Energy Efficiency and Global Warming," 2007.
13. Faruqui, Hledik, and Davis, "Sizing Up the Smart Grid," 2009.
14. See ACEEE's "Savings Estimates for ACESA," 2009(a).
15. Although EIA is prohibited from assuming policy changes in its forecasts, it is allowed to forecast the impacts of policies if asked by a member of Congress. That's what happened here.
16. See Joskow, "Regulatory Failure, Regulatory Reform" 1989: 125–200; and Fox-Penner, "Allowing for Regulation," 1988.

The Aluminum Sky

1. Miller, "The Future of the Grid," Testimony to U.S. House of Representatives, 2009.
2. For an accessible technical explanation of how transmission systems work, see my 1990 booklet, *Electric Power Transmission and Wheeling: A Technical Primer.*

3. For more technical discussions of the role of DC lines, see Peter Hartley's presentation "HVDC Transmission: Part of the Energy Solution?" Economics Department & James A. Baker III Institute for Public Policy, Rice University, and "Life Cycle 2007" by the Connecticut Siting Council.

4. See, for example, King, "Transforming the Electric Grid," 2005; Silberglitt, Ettedgui, and Hove, *Strengthening the Grid*, 2002.

5. Officially, at least, new lines in the final plan don't get any preferential treatment in their siting or environmental applications, and they sometimes aren't even liked by state utility commissions, who might have preferred another option.

6. Two early examples are the Joint Coordinated System Plan, an exercise by ten major transmission owners or RTOs in the eastern United States (http://www.JCSPStudy .org), and Osborn and Zhou, "Transmission Plan Based," 2008. At this point, these studies are purely experimental and advisory, so they have no official weight in planning or siting. They also do not come close to proposing cost allocation for new facilities, which would undoubtedly be contentious.

   The Western Renewable Energy Zone project is perhaps the most advanced regional planning process that integrates least-cost economics and is intended to proceed from study to plan to concrete line approvals. However, the process does not displace full-scale resource planning at each major western utility. See Savage, "Western Renewable Energy Zone," 2009, and related information available at http://www.westgov.org/wga/initiatives/wrez/.

7. Under the new process, the Department of Energy will select a single agency as the point of contact for all federal permits for a proposed project. All other agencies issuing approvals must conduct their approval processes concurrently and coordinate with the lead agency. See "Memorandum of Understanding among the U.S. Department of Agriculture, Department of Commerce, Department of Defense, Department of Energy, Environmental Protection Agency, the Council on Environmental Quality, the Federal Energy Regulatory Commission, the Advisory Council on Historic Preservation, and Department of the Interior, Regarding Coordination in Federal Agency Review of Electric Transmission Facilities on Federal Land," October 28, 2009.

8. A recent article in the trade publication *Megawatt Daily*, Tiernan examined the comments of many parties toward the current transmission plans of large utilities. Under the banner "Cost Allocation Clarity Needed, Parties Say," the paper reported that, "Regional transmission planning is a worthy goal to pursue in all parts of the nation, but until cost allocation issues are resolved the best laid plans may not get very far, a couple of different entities have told the Federal Energy Regulatory Commission. In comments about Entergy's revised Attachment K filing with FERC on its open-access transmission tariff, a group of municipal utilities in the utility's territory had several qualms with Entergy's regional transmission planning process, including that Entergy has not entered into "seams" agreements with adjacent

regions that would establish the framework for allocating costs of transmission projects that affect both regions. Rather than highlight the positives of the transmission planning process, Entergy's filing to comply with Order 890 and FERC's September ruling for many utilities to file revised Attachment K plans shows a lack of regional coordination on Entergy's part, the munis told FERC last week. Entergy, as well as Southern Company Services and others, made their filing last month, with Entergy and Southern highlighting the Southeast Inter-Regional Participation Process. But without clear cost allocation rules, the SIRPP effort will not produce any tangible results to improve transmission planning in the region, said the municipal systems, which included Lafayette Utilities System, the Municipal Energy Agency of Mississippi and the Louisiana Energy and Power Authority."

"That will continue to be the case until the commission addresses and resolves, once and for all, inter-system and interregional cost allocation issues and similar barriers to needed transmission development," said the munis. In the Northwest, the American Wind Energy Association and the Renewable Northwest Project had similar comments on the transmission plans of Avista, Puget Sound Energy and Bonneville Power Administration. The utilities' Attachment K filings are inconsistent with FERC's requirements because they do not provide clear, upfront cost allocation methodologies and instead refer to general guidelines adopted as part of the ColumbiaGrid planning process, AWEA and RNP said. Investors need certainty on cost allocation methodologies in order to make decisions about which transmission projects to pursue, the renewable interests said." *Megawatt Daily*, January 14, 2009.

9. The umbrella organization is known as the Organization of MISO States. Its cost allocation discussion group is known as the Cost Allocation and Regional Planning project (CARP), and the subregional effort is known as the Upper Midwest Transmission Development Initiative (UMTDI).

10. To cite just a few, the Western Governors' Association is studying renewable energy development potential and transmission expansion in the entire western grid; four subregions have their own planning efforts as well. In New England the states belonging to the wholesale markets run by ISO–New England have the New England States Committee on Electricity (NESCOE). The mid-South's efforts run out of the Southwest Power Pool and in the Pacific Northwest the Northwest Power Planning Council has a long history of regional electric expansion and energy efficiency planning.

11. EIA, Table 16, 2009(c)

12. FERC, "Electric Market Overview," 2009.

13. To view NREL's renewable resource maps visit http://www.nrel.gov/renewable_resources/. Interestingly, Shalini Vajjhala, formerly at Resources for the Future, examined NREL's data and concluded that, "Contrary to popular belief that certain types of renewable energy exist only in specific geographic regions, renewable energy po-

tential is widespread. Different states have different amounts and combinations of resources, but 43 out of 48 states have above-average potential for at least one or more renewable resources. This is promising. Few states are completely handicapped by a lack of potential resources." However, Vajjhala concedes that her analysis does not discriminate between the highly desirable utility-scale sites and less concentrated resources, and also does not systematically measure siting difficulties. See Vajjhala, "Siting Renewables," 2006: 13.

14. NERC, "Accommodating High Levels," 2009: ii [emphasis in original]; and "Integrating Locationally-Constrained Resources," 2008.

15. See, for example, NERC "2008 Long-Term Reliability Assessment," and DOEs "20% Wind Energy by 2030" (2008b).

16. A list of major new renewables line proposals as of 2009 is found in Fox-Penner, "U.S. Transmission Investment," 2009. The Edison Electric Institute compiles a utility-by-utility list of projects of all sizes in "Transmission Projects Supporting Renewable Resources," 2009.

17. To be fair, this plan was only conceptual, and it makes maximum use of all existing extra-high-voltage (765,000-volt) lines already in existence. To complete the network show, AEP estimates that only 19,000 additional miles would be needed, which is fully in line with my estimate of 20,000 to 40,000 miles. See AEP, "Interstate Transmission Vision."

18. For discussions supporting the idea, see Wood and Church, "Building the 21st Century Transmission," 2009.

19. "Joint Coordinated System Plan 2008," Available at http://www.JCSPStudy.org; Vajjhala et al., "Green Corridors," 2008; Savage, "Western Renewable Energy Zone," 2009, and DOE "20% Wind Energy by 2030," 2008(b).

20. Longer elaborations of environmentalists' concerns are found in Miller, "The Future of the Grid," 2009, and Sassoon, "Transmission Superhighway on Track," 2009.

21. For some interesting introductory discussions, see Yeager, "Congress, Think Small," 2009; Talukdar et al., "Cascading Failures: Survival versus Prevention," 2003: 25–31; Amin, "For the Good of the Grid," 2008: 48–59; and Apt et al., "Electrical Blackouts: A Systemic Problem," 2004.

22. Following a good ten years of failed efforts, Congress added a small bit of federal siting authority to omnibus energy legislation passed in 2005 (now Section 216 (h) of the Federal Power Act, a summary is available on the Department of Energy's Web site: http://www.oe.energy.gov/DocumentsandMedia/Summary_of_216_h__ rules_clean(1).pdf). The FERC was allowed to study and designate certain new transmission line routes ("corridors") that it found important to the national interest. If a line was proposed on one of these corridors, and state authorities failed to act on the application for a year, the FERC could approve the line and even condemn the land.

This "backstop siting authority" is universally viewed as too narrow and weak to make an impact. It is despised by many governors and state regulators, especially in the eastern United States, and was weakened substantially by a recent court challenge. (See *Piedmont Environmental Council v. FERC*).

The Great Power Shift

1. Platt, *The Electric City*, 1991: 76.
2. World Coal Institute, "Key Elements of a Post-2012," 2007.
3. EPRI, "Advanced Coal Power System," 2008(a).
4. EIA, "Assumptions to the AEO," 2009(f).
5. Proven U.S. natural gas reserves are about 10 years worth of annual gas demand, or about 204 trillion cubic feet (TCF), excluding supplies from Canada and Mexico. However, as pointed out to me by MIT gas expert Melanie Kenderdine, the gas supply industry has maintained a ten-year supply of proved reserves for twenty years while extracting about 600 TCF in the period. In other words, the supply industry has been able to serve all demand plus maintain a ten-year supply cushion for three decades.

   Estimates of total gas reserves, which include gas that is not yet proven to be economically recoverable, yield more than 80 years' supply—2074 TCF. "Potential Supply of Natural Gas in the United States," report by Potential Gas Committee, Colorado School of Mines, December 31, 2008, and Smead and Pickering, "North American Natural Gas," 2008.
6. *Bit Tooth Energy*, a blog written by Professor David Summers of the University of Missouri covers developments in unconventional gas drilling (http://www.bittooth.blogspot.com); also see Mall, Buccino, and Nichols, "Drilling Down: Protecting Western," October 2007.
7. See "Generators and ISO New England," 2004: 4; and "ISO New England Probes," 2004. In recent comments to the grid reliability regulator, the American Public Power Association called the industry's heavy reliance on gas power "the most immediate threat to reliability." Also see "APPA Comments to NERC," 2008. England has similar concerns, see "Dark Days Ahead," from the August 6, 2009, *Economist*.
8. The status of these technologies is closely tracked by the U.S. Department of Energy Office of Fossil Energy, the Electric Power Research Institute, and the MIT Energy Initiative, among other groups. For a sample of their recent works, see EPRI, "Advanced Coal Power Systems," 2008(a), and "The Future of Coal," 2007. As this goes to press, the U.S. Department of Energy's largest CCS project, known as FutureGen, is a 275 MW IGCC plant that faces a funding shortfall of an estimated $1 billion. The project describes itself as the first facility to integrate advanced technologies for coal gasification, electricity production, emissions control, CO_2 capture and perma-

nent storage, and hydrogen production at a commercial scale. See http://www .futuregenalliance.org/about.stm (7/3/09), and Beattie, "Thanks, But No Thanks," 2009. DOE also recently announced $408 million in funding for two CCS demonstration plants, an IGCC plant that will sequester captured CO_2 by injecting it into a depleted oil well and a 120-MW North Dakota power plant that will demonstrate a new process to capture CO_2 from coal plant exhaust gases. *Energy Daily*, July 2, 2009.

9. See the Gasification Technologies Council, http://www.gasification.org.

10. "A Milestone for Cleaner Coal," 2009. In addition to DOE's FutureGEN, the American Electric Power Company, the Southern Company, the Erora Group, Summit Power, Tanaska, and BP all have proposed IGCC projects. See Biello, "How Fast Can Carbon," 2009.

11. Friedman, "China: A Sea Change," 2009.

12. DOE, "Secretary Chu Announces," 2009.

13. For an excellent discussion of CCS retrofits, see "Retrofitting of Coal-Fired," 2009.

14. Metz and others, eds., *Carbon Dioxide Capture and Storage*, 2005: 34; also see Duncan, "Carbon Sequestration Risks," 2009.

15. The World Coal Council estimates that current sequestration projects cost as little as $6/ton, though this applies to a small number of pilot sites and not a full-scale commercial sequestration network. See World Coal Institute, "Investing in CCS," 2009(a).

    For other estimates see Biello, "How Fast Can Carbon," 2009; Sequestration capacity estimates are found in the following sources: DOE, "Carbon Sequestration Atlas," 2008; EIA "Nuclear Energy," 2009(j); and "The Future of Coal," 2007: 45.

16. A recent report from the Department of the Interior summarizes some of the high-level regulatory issues:

> First, a proposed regulatory framework must recognize carbon dioxide (CO_2) as a commodity, resource, contaminant, waste, or pollutant. Unlike most other resources that are managed, CO_2 is a material that is either being stored for disposal or is extracted for use. CO_2 is currently leased under the Mineral Leasing Act (MLA) for uses such as refrigeration (in its solid form as dry ice), fire extinguishers, and carbonation of water and soft drinks. CO_2 also is used to enhance oil recovery, which to some extent results in its sequestration. It is also important to recognize that any discussion addressing the geologic sequestration of "carbon dioxide" must distinguish between pure CO_2 and CO_2 mixed with other gases such as hydrogen sulfide, carbon monoxide, methane, and oxides of nitrogen and sulfur. These impurities have the potential to impact the economics, technical feasibility, location preferences, land use planning requirements, environmental impact mitigation, multiple-resource conflict potential, and regulatory oversight of geologic CO_2

sequestration. Impurities in CO_2 impact is value as a commodity, as well as its behavior in storage.

Second, carbon sequestration may potentially conflict with other land uses including existing and future mines, oil and gas fields, coal resources, geothermal fields, and drinking water sources. For example, sequestration in a formation would limit all future possibility of extracting minerals from the formation without some risk of venting the captured CO_2. Carbon sequestration could also have potential impacts on other surface land uses and programs such as recreation, grazing, cultural resource protection, and community growth and development. These impacts need to be addressed.

Third, a proposed statutory and regulatory framework must recognize the long-term liability of any permitting decision to sequester CO_2 and the required commitment for stewardship of facilities over an extended period of time. The scope of liability and term of stewardship will be among the longest ever attempted, lasting up to thousands of years or more. This may prove to be a potential limiting factor for siting, transportation, processing, and storage on Federal lands given the Bureau of Land Management's (BLM) multiple-use mission for long-term management of the public lands.

Many existing Federal statutes and regulations potentially apply to some aspect of the management of geologic sequestration of CO_2. These include management of other resources, waste disposal, groundwater protection, and human health and safety. However, due in part to the many unique challenges discussed above, gaps may exist in the current laws and limit our ability to address the range of circumstances, scope of potential liability, required timeframe of stewardship, and regulatory primacy differences between the states and Federal Government. (U.S. Department of the Interior "Framework for Geological Carbon," 2009(a))

As of this writing, EPA has proposed draft rules to regulate CO_2 injections under the Safe Drinking Water Act. The proposed rule, issued in 2008 after three years of study, runs 221 pages and covers everything from conducting and filing initial geological surveys to financial conditions the sequestering firm must meet, such as having sufficient insurance in place. See EPA "Federal Requirements under the Underground," 2008.

17. Gronewold, "N.Y.'s Pioneering Effort," 2009.
18. Deutch and others, "Update of the MIT 2003," 2009.
19. Cohen, Fowler, and Waltzer, "'NowGen': Getting Real," 2009.
20. See "Meeting Projected Coal Production," 2008; and Smith, "U.S. Foresees," 2009, for recent discussions on the topic, and watch for reassessments of coal reserves from the U.S. Geological Survey and EIA. For discussions of environmental concerns, see http://www.epa.gov/owow/wetlands/guidance/mining.html.

21. "Nuclear Power in the USA," 2009. While the majority of plants were built during this time, any further investment during this time came to a halt due safety and cost concerns.

22. Jaczko, 2009.

23. See "Nuclear Energy Outlook 2008," 2008; and Goering, "Nuclear Plants Being Revived Worldwide," 2009.

24. See Alexander, "Blueprint for 100 New Nuclear," 2009.

25. The availability and cost of nuclear fuel does not seem to be a problem; there are ample deposits of uranium around the world. However, nuclear plants use large amounts of cool water and must shut down for safety reasons if their water supply becomes too warm. Since 2006, heat waves have caused water-related shutdowns in several plants in the southern United States and threatened many others; the summer 2008 heat wave reportedly closed as much as one-third of France's nuclear capacity. See Pagnamenta, "France Imports UK Electricity," 2009; EDF, "Achievements: Controlling Human Impact on the Environment," 2007; and Fleischauer, "Heat Wave Shutdown," 2007.

26. Beattie, "Utilities Unprepared," 2009.

27. The translation of nuclear power's business risks are discussed by the major utility bond rating agencies: Moody's and Standard & Poor's (for an example, see Hempstead, "New Nuclear Generating Capacity," 2008; and "Nuclear Plant Construction," 2008). These risks are also discussed in Severance's, "Business Risks to Utilities," 2009.

28. Congressional Budget Office, "Nuclear Power's Role," 2008; and Cooper, "The Economics of Nuclear Reactors," 2009.

29. See Romm, "The Staggering Cost," 2009(a); and "Warning to Taxpayers, Investors," Center for American Progress, January 7, 2009(b).

30. The IEEE reports that there are only four nuclear power plant manufacturers worldwide. See McClure, "Energy Fixes: Smart Grid," 2008.

31. See Toole and Winsor, "Nuclear Workforce Dwindles," 2009; Gellatly, "Study: Nuclear Industry Needs 10K," 2009; "Readiness of the U.S. Nuclear," June 2008; IAEA, "Nuclear Power Industry's Ageing," 2004; and "Draft: Workforce Planning," 2009.

32. The shortages are not only in the white collar workforce. Many construction tradespeople on nuclear projects, such as welders, require advanced training and certification. The supplies of these workers cannot be rapidly expanded—it takes time to locate and train these workers. Two years ago, I heard a representative from a worldwide nuclear engineering firm tell a conference that he estimated that building one hundred nuclear plants in the United States in the next twenty years would require every nuclear-certified welder in the world. Specialized training and R&D facilities are also needed. See "Nuclear Energy for the Future," 2008, for a recent catalog of the industry's R&D needs. Conversely, former Nuclear Regulatory Commission

(NRC) chairman Richard Meserve recently noted to me in an email that enrollment in nuclear engineering programs has started to increase.

33. Spacing of turbines is determined by the diameter of the turbine rotors, which average around 70 meters for a 1.6 MW unit. Spaced with 10 diameters between rows and 3 to 5 diameters per turbine, the average utility-scale wind farm requires 30 to 60 acres, not including infrastructure. For a longer discussion on the spacing of wind turbines see: "Wind Power Project Site," 2005; "Wind Farm Area Calculator," National Renewable Energy Laboratory, available at: http://www.nrel.gov/analysis/ power_databook/calc_wind.php; and "Wind Web Tutorial," American Wind Energy Association, available at: http://www.awea.org/faq/.

34. Wiser and Bolinger, "2008 Wind Technologies," July 2009.

35. EIA shows wind farm overnight construction costs dropping from $1923/kW to $1615 by 2030, a reduction of 16%. While U.S. DOE, Energy Efficiency and Renewable Energy's (EERE's) highly touted 20%-by-2030 report "20% Wind Energy by 2030," 2008(a), expresses some doubt over the higher end of the range, it is clear that manufacturers and installers have more innovation in the pipeline, such as carbon composite blades and new construction techniques. Also see Neville, "Prevailing Winds," 2008. Most other elements of wind power costs have also been declining, but rising commodity costs caused wind turbines to become more rather than less expensive during the last five years. See U.S. DOE, "Cost Trends."

36. DOE, EERE, "20% Wind Energy," 2008(b).

37. There are large-scale storage technologies that help regulate (balance) the grid over very short time scales (i.e., smoothing out power over a period of literally a few seconds). These storage technologies are very promising and likely to see much greater use in the high-voltage grid. However, these technologies do not balance load and generation on the time scale shown in Figure 8-1. See "Bottling Electricity," 2008; and Gyuk, "Energy Storage Applications," 2008.

38. It is also worth noting that wind power's economics seems to work best in the large, centralized, competitive wholesale markets, where each hour of output is priced separately. This argument is laid out carefully in "Facilitating Wind Development," by Kirby and Milligan, 2008: 40–54, and is advanced by advocates for expanding these wholesale markets nationwide, such as the COMPETE coalition (http://www .competecoalition.com).

39. Many utilities have conducted studies of the costs of wind integration; two experts at a Department of Energy laboratory summarized them as of a year ago and found an average short-term integration cost of about 0.1 cents per kilowatt-hour, see Wiser and Bolinger, "Annual Report on U.S. Wind," 2008. A major study of this question is due in October 2009 from the Federal Energy Regulatory Commission as well. For an excellent nontechnical overview, see "Integrating Wind Power," 2009.

40. Despite reports of high costs for offshore wind projects emanating from Europe (e.g., Hansford and Rowson, "Capital Costs Undermine Offshore," 2009; and Fair-

ley, "Germany's Green-Energy Gap," 2009, interest in developing offshore wind remains intense among governors and other economic development officials and groups in coastal states. The governors of ten states on the eastern seaboard recently sent a joint letter to the U.S. Congress opposing a "transmission superhighway" from the wind-rich Dakotas to the East Coast, arguing that they preferred to develop their own offshore wind.

41. Strictly speaking there are three types of crystalline silicon panels: single crystal, polycrystalline, and ribbon; the first and most common variety are blue-black.

42. NREL, "Renewable Energy Cost Trends," 2005.

43. See Pernick and Wilder, "Utility Solar Assessment (USA) Study," 2008. These costs do not include storage, backup, or integration.

44. "Increasing Efficiency," Mitsubishi Electric Corporation.

45. Hand, "PV Materials," 2009.

46. Koning, "Renewable Energy: Feeling the Heat," 2009.

47. Wiser, Barbose, and Peterman, "Tracking the Sun," 2009: Figure 5.

48. For more detail, see Appendix B, Part 1, Summary of Selected Large-Scale Power Generating Technologies. This table projects possible changes in the costs of large-scale generating technologies up to 2030. For an interesting discussion of how BOS costs might decline, see Newman and others, "Accelerating Solar Power Adoption," 2009.

49. Sherwood, "U.S. Solar Market Trends 2008," 2009. Many other utilities announced plans for multi-megawatt PV installations; Pacific Gas and Electric, the San Francisco–based utility, shattered all records by announcing a plan to build a record-breaking 550-MW PV plant in California. SEIA, "U.S. Solar Industry Year," 2009.

50. SEIA, 2009: For good descriptions of the two technologies, see the Web sites of the National Renewable Energy Laboratory and the DOE Office of Energy Efficiency and Renewable Energy.

51. "PG&E and BrightSource," 2009. Abengoa Solar has an 11 MW and a 20 MW solar power tower operating in Seville, Spain, but there is little sign of interest in developing this technology commercially as of now in the United States. http://www .abengoasolar.com/sites/solar/en/our_projects/solucar/ps10/index.html.

52. U.S. Department of Interior "Secretary Salazar, Senator Reid Announce," 2009(b).

53. Kelly, "Large Plant Studies," 2006.

54. The DOE estimates that dry cooling adds about 10% to the cost of power from a CSP plant; estimates from the Electric Power Research Institute seem to be higher. See Price, "Cooling for Parabolic," 2008: 19.

55. EIA, "Assumptions to the AEO," Table 8.13, 2009(f).

56. Some forestry practices may also produce zero carbon biomass. The Food, Conservation, and Energy Act of 2008 (FCEA), known widely as the 2008 Farm Bill, recognized a range of forestry activities for eligible carbon-offsets such as afforestation, reforestation, conservation, forest management, and harvested wood products. In a

letter from utilities and trade associations encouraging the inclusion of forestry biomass in Waxman–Markey RES and offsets, it was estimated carbon-neutral biomass has the potential to provide as much as one-third of the proposed Renewable Energy Standards. (Available at: http://www.nafoalliance.org/LinkClick.aspx?file ticket=03VDdh6B0ws%3d&tabid=65&mid=510).

57. All but four of the 190 U.S. biomass power plants are owned by private firms with access to cheap fuel and independent power producers who contract for fuel. See EIA, "Biomass Milestones," and EIA, "Renewable Energy—Biomass Data and Information: Biomass, Wood/Wood Waste, and Municipal Solid Waste." Available at: http://www.eia.doe.gov/fuelrenewable.html. With the onset of carbon limits, utilities are showing much more interest in buying from or building biopower plants.

58. See Slack, "U.S. Geothermal Power Production," 2008; and U.S. DOE, "Renewable Energy Data Book," 2008(b).

59. Because geothermal plant sites average 50 MW, they are not large enough to justify long dedicated new transmission lines. This is a somewhat unique constraining factor.

60. A 2007 report by the Geothermal Energy Association calls geothermal energy "an underestimated, underreported, underexplored, and understudied natural resource that could have a large impact on America's future energy supply." The report cites near-term resource estimates of 24,000 MW from the National Renewable Energy Laboratory, primarily in California. See Fleischmann, "An Assessment of Geothermal," 2007. Also see, U.S. Geological Survey "Assessment of Moderate- and High," 2008; Bertani, "Long-Term Projections," 2009; and Fridleifsson and others, "Possible Role and Contribution," 2008. Google.org has also sponsored substantial research into this energy resource.

61. See Schwartz, "First Commercial Hydrokinetic Turbine," 2008. For an excellent short overview of "damless" hydroprojects in the public power sector, see Hequet, "Damless," 2008.

62. Lobsenz, "UK Moves Forward," 2009.

63. Sims, "Hydropower, Geothermal, and Ocean Energy," 2008: 389–395.

64. Rawson and Sugar, "Distributed Generation and Cogeneration," 2007: 1.

65. There are other important small-scale generation technologies I do not discuss here purely to make the discussion more manageable. These include small-scale biomass generators, which are significant in some DG markets (e.g., Appendix B, Part 2) and small-scale hydropower. There are thousands of small hydroelectric dams in the United States that can be upgraded with better technology, and many excellent new small hydrotechnologies coming into the marketplace, but it is difficult to tell how large a contribution small hydro will make.

The most common small-scale generator in the United States by far is the diesel engine-sets. These units are plentiful because they have long been the cheapest small onsite generator. They can run on natural gas and other fuels, though diesel is most common. Since nearly every large building in the United States is required to have an emergency generator, there are an estimated 50,000 to 75,000 MW of these engines sitting idle in basements and parking lots around the country. See Rocky Mountain Institute "Micropower Database," 2008; and "Market Forecasts," Distributed-Generation.com. Available at: http://www.distributed-generation.com/market_forecasts.htm.

Although they are cheap and reliable, they do not make power as efficiently as natural gas combined cycle power plants, even after you include the costs of transmission and distribution. (However, these units can be used much more efficiently for cogeneration, which we'll discuss in a moment). They produce much more air pollution per kilowatt-hour, including twice as much CO_2 per kWh as natural gas power plants. Worse still, they emit these pollutants in the worst possible locations, mainly at ground level sites in the middle of congested urban areas. The environmental case against increased reliance on diesel generators is elaborated in Greene and Hammerschlag, "Small and Clean Is Beautiful," June 2000: 50–60. The authors note, for example, that a 0.5% increase in generation from small diesel engines would increase total nitrogen oxide pollution by 5% nationwide because these engines have no emissions controls. For all these reasons, hardly anyone in the industry thinks diesel engines should become a major contributor to the Smart Grid, and they don't appear in the Table 8-1.

66. Electricity-displacing renewable energy technologies are recognized as a technology group by the California Solar Initiative (see CPUC, California Solar Initiative, *Program Handbook*, 2009: 19, Sec. 2.2.3).

67. In 1978 Congress enacted federally mandated regulatory reforms that forced utilities to interconnect with cogenerators and pay them the utilities' "avoided cost," the complicated regulatorily determined rates I have cautioned about several times already. These reforms created a large wave of new cogeneration plants in the 1980s, but the wave tapered off when the cost of gas and power declined in the 1990s. Also, power plant owners figured out a way to use their own waste heat to make more power; these are the gas combined cycle power plants we met at the start of the chapter.

68. Although Table 8.1 shows only two size categories, the National Renewable Energy Laboratory uses three size classes—residential (under 30 kilowatts, the peak load of one large house); medium (30 to 500 kW); and commercial scale (500 kW to 2 MW or more per turbine). See "Wind Power: Wind Technology Today," University of Massachusetts at Amherst. For brevity, my table omits the middle category.

69. "AWEA Small Wind Turbine Global Market Study: 2009."

70. Ibid.: 3; and Forsyth, "Small (Distributed) Wind Technology," 2008.

71. Romm, *The Hype about Hydrogen*, (2004, 2005), explains why fuel cells are unlikely to become a dominant energy source.

72. For another similar assessment see Figure 2.10 in the National Research Council's 2009 Report, *America's Energy Future*. The Council's estimates do not include renewable credits, as do the 2010 estimates in my table, and are therefore higher for renewable sources. Otherwise, the two sets of estimates are similar. See Appendix B for further details and a graphical comparison of the two sets of estimates.

Billion Dollar Bets

1. U.S. lawmakers are aware of the problem of excessive volatility and are proposing measures that will reduce it. These include allowing companies to bank their allowances for use in one of the next several years and borrowing a limited number of allowances from future years' allocations. Other limits on volatility under discussion are limits on allowance prices or price changes or the creation of allowance reserves that can be released if market prices become too high. For a discussion of volatility reduction strategies in Waxman–Markey American Clean Energy and Security Act of 2009, see National Commission on Energy Policy, "Forging the Climate Consensus," 2009. For a discussion of additional strategies using the Bingaman–Specter Low Carbon Economy Act of 2007, see U.S. Senate, "Bingaman-Specter Low Carbon Economy Act."

2. For a longer discussion and analysis, see Celebi and Graves, "CO_2 Price Volatility," 2009; and Chupka, "Uncertainty, Volatility and Risk," 2008. There are also plenty of carbon-related risks other than allowance prices, such as the chance that carbon policies will change or that low-carbon generating technologies won't work as planned.

3. "Scenarios: Shooting the Rapids." *Harvard Business Review*, Nov–Dec 1985, no. 85617. Wikipedia puts it nicely as well: "Above all, scenario planning is a tool for collective learning, reframing perceptions and preserving uncertainty when the latter is pervasive. Too many decision makers want to bet on one future scenario, falling prey to the seductive temptation of trying to predict the future rather than to entertain multiple futures. Another trap is to take the scenarios too literally as though they were static beacons that map out a fixed future. In actuality, their aim is to bound the future but in a flexible way that permits learning and adjustment as the future unfolds." Available at: http://en.wikipedia.org/wiki/Scenario_planning.

4. See Google, "Clean Energy 2030," 2008; and Sovacool and Watts, "Going Completely Renewable," 2009.

5. See EIA, "Existing Capacity by Producer Type," 2009(d); Shipley and others, "Combined Heat and Power," 2008; and Hedman, "CHP Market Status," 2005: 8.

6. For a selection of potential estimates, see Shipley and others, "Combined Heat and Power," 2008; Committee Workshop on Combined Heat and Power, 2009; "Cooling, Heating, and Power," 2003; "Combined Heat and Power Market," 2004; Cleetus,

Clemmer, and Friedman, "Climate 2030," 2009; ACORE, "Outlook on Renewable Energy," 2007; Casten, "CHP: One of the Answers" 2009 (also for an industry perspective); and "Sector Profiles of Significant Large CHP Markets," 2004.

7. The U.S. has experienced some success encouraging CHP, albeit at a high price. In 1978 Congress passed the Public Utility Regulatory Policies Act (PURPA), which required distribution utilities to interconnect with "qualified" CHP plants and buy their surplus power. The rate for purchasing the surplus was required to be that distributors avoided cost of generation, set by regulators if the utility was state regulated. This is essentially the same avoided cost discussed in Chapter 5, with its many measurements challenges. Some state regulators forced utilities to sign long-term contracts with CHP plants to purchase at avoided cost rates that greatly overestimated long-term avoided costs. This led to the de facto abandonment of this approach and formal repeal of PURPA in areas with fully developed wholesale power markets. In the period between enactment in 1978 and de facto repeal in 1996, CHP grew from 10,000 MW to 43,000 MW. (For additional discussion, see Fox-Penner, "Impact of the Public Utility," 1996.)

Both the Obama stimulus legislation and the proposed Waxman–Markey bill provide substantial new support for CHP, primarily in the form of tax credits. See U.S. Census "Summary of Energy Efficiency," and United States Clean Heat & Power Association, "Clean Heat."

8. I should also note that microturbines look like they will become increasingly cost-competitive, and if natural gas prices stay reasonable there is every reason to believe that they will steadily gain market share in residential and commercial buildings. Their growth will be constrained by the inevitable delays in adopting any new technology (even one with tax credits) and the pace of new building construction, since these are much easier to install in new construction rather than to retrofit them into existing buildings.

9. Rawson and Sugar, "Distributed Generation and Cogeneration," 2007.

10. Ibid., 20. Unfortunately, the latest prognosis is even worse. Noting that it is "faced with the slow development of new CHP in California," the CEC launched a workshop to improve its CHP policies further, see Committee Workshop on Combined Heat, 2009.

11. NEI, "Policies That Support New Nuclear Development," 2009.

12. Ibid., and U.S. Congress, "American Clean Energy and Security," 2009: 86–104.

13. "Carbon Capture and Geological Storage," 2009.

14. In a recent survey of nuclear's prospects, a world-class team of MIT engineering professors gave this downcast assessment: "The sober warning is that if more is not done, nuclear power will diminish as a practical and timely option for deployment at a scale that would constitute a material contribution to climate change risk mitigation" (Deutch and others, "Update of the MIT 2003 Future of Nuclear Power," 2009).

The Organisation for Economic Co-Operation and Development's (OECD's) 2008 Nuclear Energy Outlook puts a more positive slant on nuclear's prospects, but its conclusions reinforce the remote chance that nuclear will *displace* any of the alternatives. Under its high-nuclear scenario, which assumes that all of the risk factors fall in favor of nuclear and against coal and renewables, nuclear power would supply 22% of the world's electricity by 2050—about the same fraction it supplies today in the United States. Obviously, over three-fourths of world power would come from coal, gas, and renewables—and probably more in the United States. For more skeptical views, see the works of Joe Romm (http://www.climateprogress.org) and Bradford, "The Myth(s) of the Nuclear Renaissance," 2009.

15. DOE, "20% Wind Energy," 2008(b); and Baxter, "Energy Storage," 2008: 102–112.
16. California Energy Commission, "Renewable Energy Transmission Initiative," 2009.
17. Douglas and others, "33% Renewables Portfolio Standard," 2009.
18. "Strategies to meet possible 33%," 2009.
19. Petrill, "Creating a Secure Low," 2008; Weyant, "EMF Briefing U.S. Climate," 2009; Romm, "Is 450 ppm (or Less) Possible?" 2009; Marshall, "Policy: Carbon Capture and Storage," 2009; EPA, "EPA Analysis of the American," 2009; and Google, "Clean Energy 2030," 2008. The Energy Modeling Forum is actually a compendium of results from prominent energy sector models, each with different structures and different assumptions. Of the six models reviewed in EMF 22, one had very little CCS coal and a second eliminated nuclear. Other studies of climate options that include the full portfolio of options include EIA, "Energy Market and Economic Impacts," 2009(k); Pacala and Socolow, "Stabilization Wedges," 2004; IEA, "Energy Technology Perspectives," 2008(a); Sullivan and others, "Comparative Analysis of Three," 2009; and Duke and Lashof, "The New Energy Economy," 2008.
20. "Cutting CO_2 emissions," 2009.
21. Marshall, "Policy: Carbon Capture and Storage Moves," 2009. Joe Romm also comments on this point in "Is 450 ppm (or less) possible?" 2009.
22. Sioshansi, "Carbon Constrained," 2009: 73.
23. Chupka and others, "Transforming America's Power Industry," 2008.

Energy Efficiency: The Buck Stops Where?

1. "Behavioural economics is concerned with the empirical validity of these neoclassical assumptions about human behaviours and, where they prove invalid, with discovering the empirical laws that describe behaviour correctly and as accurately as possible. As a second item on its agenda, behavioural economics is concerned with drawing out the implications, for the operations of the economic system and its institutions and for the public policy, of departures of actual behaviour from the neoclassical assumptions." *The New Palgrave; A Dictionary of Economics*, Eatwell, Milgate, and Newman, eds. (London: Macmillan, 1987). Vol. 1, A to D: 221. A very

interesting set of papers is in the proceedings of the 2008 Behavior, Energy, and Climate Conference, the Precourt Center for Energy Efficiency, Stanford University: Available at: http//piee.standfor.edu/cgi-bin/htm/Behavior/2008_becc_conference _online_progam.php.

2. Thaler and Sunstein. *Nudge*, 2008: 6–7.

3. Ibid.: 9

4. The boost in employment levels is tied to the fact that efficiency options often require many more labor-hours per dollar spent in comparison to the construction of power plants. While much of the material and manufacturing of power plants now comes from outside the United States, most EE projects have a high proportion of their budgets spent in the United States, creating direct and indirect employment. Recent studies include Bezdek, "Renewable Energy and Energy," 2007. For additional estimates see Pollin and others "Green Recovery," 2008; and Pinderhughes, "Green Collar Jobs," 2007.

5. Levine and Ürge-Vorsatz, "IPCC WGIII Assessment Report," 2007. Also see Figures SPM9, SPM10 in "Climate Change 2007," 2007.

6. McKinsey & Company, "Reducing U.S. Greenhouse Emissions 2007; Nadel, "Energy Policy Trends," 2009(a); Cleetus, Clemmer, and Friedman, "Climate 2030," 2009; and IEA, "Energy Efficiency Policy Recommendations," 2007.

7. See Claussen, "Energy Efficiency, Climate Change," 2004.

8. Gillingham, Newell, and Palmer. "Energy Efficiency Economics and Policy," 2009.

9. Sathaye and Phadke. "Representing Energy Demand," 2008.

10. Faruqui, "Will the Smart Grid Promote," 2008. Also see, Parmesano, "Rate Design Is the No. 1," 2007: 18–25; and Kiesling, "Project Energy Code—Markets," 2009, for more on why prices are important.

11. First, the inaccuracies introduced by weaker price signals often do not greatly affect the calculation of net benefits from many EE measures. Some measures, like more efficient refrigerators, save energy during all hours, so using an average price is actually correct. For other measures that save on peak, using average prices underestimates the value of EE. We aren't tapping anywhere near the full EE potential even with these underestimates.

    Another reason why average pricing shouldn't stop us is that many utilities and EE experts don't use average prices when they do cost-effectiveness calculations, even if this is what customers are charged. For example, many residential customers pay average rates of about 10 cents/kWh (although California is likely to be the first state to change entirely to time-based pricing in the next few years). However, when California utilities calculate the value of an efficiency measure, they use the forecasted cost of making power in every hour of the year from a model developed specially to ensure accurate EE calculations.

12. Jaffe, Newell, and Stavins, "Technological Change and the Environment," 2000.

13. There is a vast literature debating the validity of market barriers and public

efficiency policies. Several of the many good papers are references: "The elusive negawatt—Energy efficiency," 2008; Jaffe, Newell, and Stavins, "Energy-Efficient Technologies," 1999; Cavanagh, "Energy Efficiency in Buildings,"; Taylor and Van Doren,
"Myth Five—Price Signals," 2007; "Quantifying the Effect of Market," 2007; Sutherland, "Market Barriers to Energy-Efficiency Investments," 1991; Golove and Eto,
"Market Barriers to Energy Efficiency," 1996; Dennis, "Compatibility of Economic,"
2006: 58–72; and Brown, "Market Failures and Barriers," 2001: 1197–1207. A similar
and more principled objection comes from doubts that energy efficiency opportunities are in fact low cost when all the costs and savings are measured accurately.

14. For a very important compendium of EE policy prescriptions, see the National Ac-
    tion Plan for Energy Efficiency, 2006.
15. Meyers, McMahon, and McNeil, "Realized and Prospective Impacts," 2005.
16. deLaski, Testimony before the U.S. House, 2009.
17. ACEEE, "U.S. DOE Heats Up," 2009(b).
18. Krauss, "Tightened Codes," 2009.
19. Ibid. State-by-state information and other valuable information is on the Web site
    of the Building Code Assistance Project, http://www.bcap-energy.org.
20. U.S. Congress, Section 201 of the American Clean Energy and Security Act (2009).
21. Building Codes Assistance Project, "Residential Building," and "Commercial Build-
    ing," 2008.
22. "Solar Water Heater Rebate," Progress Energy, 2009. Available at: http://www
    .progress-energy.com/custservice/flares/save/solarheater.asp.
23. "Energy Revolving Fund: Low-Interest Loans for Energy Efficiency Improvements,"
    Missouri Department of Natural Resources. Available at: http://dnr.mo.gov/energy/
    financial/loan-information.htm.
24. Kats and Carey, "Upgrading America's Homes," 2009.
25. Johnson, "Municipal Energy Financing," 2009. Also see New Rules.org. "Municipal
    Financing for Renewables and Efficiency." Available at: http://www.newrules.org/
    energy/rules/municipal-financing-renewables-and-efficiency.
26. Kushler, York, and White, "Meeting Aggressive New State," 2009: iii.
27. See Nadel, "Success with Energy Efficiency," 2009(b); and Testimony before U.S.
    House of Representatives, 2009(c).
28. I should also note that most utilities outsource a fair amount of their actual installa-
    tion work to ESCOs, tapping their installation expertise. In a sense, utility programs
    should be viewed as utilities being in charge of marketing, managing, and capitaliz-
    ing the EE function, while ESCOs provide the implementation services.
29. "Implementing California's Loading Order," 2005. Graphics based on these data
    were generated by PG&E for use in efficiency presentations.
30. Kushler, York, and White, "Meeting Aggressive New State," 2009: 21.
31. Another somewhat smaller downside to government EE funding is that it could
    make interactions between CO_2 trading markets and EE projects a little less effec-

tive. EE projects can qualify for offsets, which are akin to tradable carbon emissions permits. If state governmetts are the primary local funders of EE projects, there may be a temptation to take back the offsets rather than let the market use them to incentivize greater savings. In the case of utility funding, it won't matter quite as much if the utility or the EE project gets the offset, as either entity will be incentivized to use the permit efficiently.

32. Nadel, Testimony before the House of Representatives, 2009(c): 11.

Two and a Half New Business Models

1. For useful references on network effects in economics, see Brennan, "Network effects as infrastructure challenges facing utilities and regulators," 2009; Shy, *The Economics of Network Industries*, 2001; and Moss, ed., *Network Access, Regulation and Antitrust*, 2005: 1–5.

2. See Coase, 1988; and Williamson, 1975, 1979.

3. For excellent formal discussions of the economics of vertical integration and network effects, see Joskow, "Vertical Integration," 2003; and Klein, "The Make-or-Buy Decision," 2004.

4. Joskow and Schmalensee, *Markets for Power*, 1988: 93.

5. See Kwoka, *Power Structure Ownership, Integration*, 1996: 141. I review the literature ten years later in Fox-Penner, *Electric Restructuring: A Guide to the Competitive Era*, 1997. More recent studies include Kuhn and Machado, "Bilateral Market Power," 2004; Finon, "Investment Risk Allocation," 2008: 150–183; Greer, "A Test of Vertical Economies," 2008: 679–687; Arocena, "Cost and Quantity," 2008: 39–58; Mansur, "Upstream Competition," 2007: 125–156; Fraquelli, Piacenz, and Davide, "Cost Savings from Generation," 2005: 289–308; and Nemoto and Goto, "Technological Externalities," 2004: 67–81. These recent studies are unanimous in their verdict that vertical integration savings are significant.

6. Michaels, "Vertical Integration and the Restructuring," 2006; Kwoka, *Power Structure Ownership, Integration*, 1996; and Faruqui, Sergici, and Wood, "Moving toward Utility-Scale," 2009; Kwoka, Ozturk, and Pollit, "Divestiture Policy," 2008; and Kwoka, "Vertical Economies in Electric Power," 2002.

7. These data come from FERC Form 1 filings collected by Ventyx and a classification of lines of business in Platts Top 250 Energy Companies 2008, directed by Dr. José Antonio García, of *The Brattle Group*'s Madrid office. Vertical integration is also proving to be even prevalent in other deregulated network industries such as telecommunications.

8. Jones et al., "Electric Utilities," 2002.

9. In his extensive survey of vertical integration across all industries, Professor Peter Klein writes:

A second lesson is that vertical relations are often subtle and complex. While early empirical work on transaction cost determinations of vertical integration tended to focus on black-and-white distinctions between "make" or "buy," researchers increasingly recognize that a wide variety of contractual and organizational options are available; there are many shades of gray. The literature on hybrids has grown dramatically in the last ten years, while there are fewer studies of mundane issues such as outsourcing versus in-house production per se.

10. Ibid.: 456.
11. EIA, 2009(i).
12. In *Power Structure Ownership, Integration*, 1996, Kwoka discusses the rare instances in which cities have multiple distribution systems.
13. Plant additions are based on Velocity Suite data from Ventyx and reflect nameplate capacity expansions as well as additions. Units less than 50 MW are omitted, emphasizing that this trend is occurring outside of the adoption of DG units (under 20MW).
14. When generation competition is introduced, it is theoretically preferable to forbid the transmission owner from owning generators. This is because access to the grid on reasonable terms is essential for competitive generators to be able to ship their power to customers. If grid owners also own generators they have incentive to deny access to generators they don't own in order to lessen the competition faced by the deregulated generators they do own. This incentive to foreclose transmission disappears when grid owners do not own any generation as they no longer care which generators are successful and which are not.

      Throughout the world two approaches have been taken to removing this incentive. In the United Kingdom and most retail choice states, grid owners are simply not allowed to own generators. Obviously, in this case there will be no vertical integration economies preserved. In the rest of the United States and much of Europe, grid owners are allowed to own generators, but their transmission operators must grant equal access to any rival generator who wants to use their system. This so-called open access transmission regime is how wholesale competition operates in the United States. The extensive system of rules and procedures created by the FERC to effectuate open access removes some but not all of the benefits of vertical integration. For information on U.S. open access, see FERC's Order No. 1000.
15. For a much more extensive bibliography of studies analyzing the impacts of wholesale and retail competition see Appendix C: Further Discussion and Reading Regarding Competition in the Power Industry.
16. Utility consultants Michael Beck and William Klun write:

      Change is Accelerating: Utilities are not renowned as incubators of change. Generally they're managed as conservative, low risk businesses. Utilities don't lead the way in the development of new technologies. R&D functions, if extant in a utility, are primarily geared around monitoring, supporting and underwriting industry organizations that undertake research for groups of companies. Given the critical nature of electricity, new technologies are introduced only after careful testing and usually in limited applications. There's no reward provided for being first to market with any development; in fact, there is more downside risk to any new technology introduction than upside benefit. Industry history, therefore, is one of careful and slow change.

17. See Compete (2009) for this and related statistics and studies. It is not clear from the historical or economic record that deregulation was the necessary or causal force, especially at retail. See the text following.
18. One of the best articulations of this view was written by then commissioner Jon Wellinghoff and David Morenoff, "Recognizing the Importance of Demand Response: The Second Half of the Wholesale Electric Market Equation," *Energy Law Journal*, 2007.
19. Galvin and Yeager, "Perfect Power," 2009; Oatman and Crudele, "The Galvin Path to Perfect Power," 2007.
20. Kiesling, "Project Energy Code," 2009.
21. This view is nicely articulated by Mark Jacobs, CEO of Reliant Energy, in a recent interview in the journal *Public Utilities Fortnightly* by Burr, June 2008: 74.

> **Fortnightly:** It's been a year since the Texas price-to-beat structure was removed. How has that affected competition in the state?
>
> **Jacobs:** In my view, today we're still in the early stages of seeing the benefits of a competitive market. Most of the competition is still price-based as opposed to value-based. When we deregulated telecom in 1984, most of the early competition was price based. All the technology we have in the palms of our hands today was unthinkable 20 years ago when we deregulated. It didn't happen right out of the chute.
>
> We're going through a similar process in power. Now as a customer you can pick a variable-rate plan, a fixed-rate plan, different tenors in terms of your contract for power supply, and more degrees of freedom. But mainly the products are differentiated around price. The next evolution in the market will move beyond differential pricing and into value added services.
>
> Today in the retail business in Texas we sell a commodity. What we're working toward is providing people with a value-added service that helps them buy that commodity. It sounds similar but it's very different. Electricity purchases in Texas are a big line item, so helping people be thoughtful about that and understand what they are buying, when they buy it, will be valuable.

> **Fortnightly:** Arguably retail deregulation didn't work in most states the first time around because meters weren't up to the task of delivering added value. Is the meter the key?
>
> **Jacobs:** The smart meter is kind of like computer hardware. By itself it doesn't do anything. We and other companies are providing the software to take advantage of the hardware. We're providing solutions that will give customers valuable benefits—such as disaggregated rates and time-of-use rates.
>
> The smart meter is the enabling technology. It will allow the power of competition to take over.

22. My thesis regarding climate policies and vertical integration is that policymakers will want assurances that utilities will meet their emissions reductions targets and as much assurance as possible regarding future prices and costs. If utilities rely primarily on the market to supply energy sources and carbon emissions reductions, providing assurances regarding long-term costs will be difficult. They will undoubtedly employ long-term contracts with fixed or formula prices, but integration has proven to be preferable to such contracts in many cases.

    I develop this line of reasoning a little further in my testimony in a Colorado regulatory proceeding, *Public Utilities Commission of the State of Colorado*, Docket No. 07A-447E, June 9, 2008. The reasons have less to do with how well retail competition works and more to do with the continuance of vertical integration and the continuing importance of climate change, which will put a premium on assurances of low-carbon supplies.

23. Carlisle, Elling, and Penney, "A Renewable Energy Community," 2008: 1–2.

The Smart Integrator

1. Theory and practice are well explained in many works by Harvard professor William Hogan, such as "Competitive Electricity Market Design: A Wholesale Primer" (December 18, 1998) (57 pages), a compilation and update of introductory materials from previous papers. Also see Stoft's, "Power System Economics," 2002, and the seminal work by Schweppe et al., *Spot Pricing of Electricity*, 1998.

2. "eData Feed FAQs," PJM Interconnection, available at: http://www/pjm.com/Home/faqs/esuite-etools/edatafeed.aspx; "PJM Bus Model as of May 13, 2009," PJM Interconnection, available at: http://www.pjm.com/markets-and-operations/energy/lmp-model-info/bus-price-model.aspx; and Reitzes and others, "Review of PJM's Market Power," 2007.

3. Markets are starting to be established for generation capacity, but they are still rather new and not fully accepted in the industry. Markets for other sources of value from DG, and for the systemic costs DG imposes on systems, are either nonexistent

or still in their infancy. All of these markets have been designed for large-scale sources. While there is no theoretical reason why they won't work at the local level, we are far from knowing how practical it will be to implement them locally.

4. Here, I differ somewhat with the deregulation proponents, who believe that the Smart Grid will unshackle the smart marketplace from regulators. My view is that regulation will remain extremely pervasive even if retail prices are allowed to "float" within regulated limits, and that state regulators will exert by far the largest influence shaping market outcomes for quite some time.

5. Vojdani, "Smart Integration," 2008.

6. Arnold and Cochrane, "Future Opportunities for the Energy," 2009. IBM's utility experts explain the large, advanced information systems Smart Grid utilities will need in Welch and McLoughlin, "Information is Power," available at: http://www-03.ibm.com/industries/utilities/us/detail/resource/Y448128V62075A69.html.

7. You may have read or heard about more sophisticated forms of rate regulation that are usually called incentive or performance-based regulation (PBR). This is a huge family of regulatory models that range from very simple performance-based adjustments to profits to complex mathematical formulas used to set rates. These are often excellent refinements on the basic cost-of-service regulation, but they don't change the fundamental incentives we are talking about here. They improve regulatory oversight, discourage wasteful spending, and they address specific incentive incompatibilities, but not the disincentives to sell less power. Nearly every regulatory agency in the United States builds some aspects of PBR into its regulatory processes, and the United Kingdom uses it exclusively to set transmission rates (it is commonly called RPI-X regulation there, a shorthand based on the rate-setting formula). For further information see Sappington and others, "State of Performance-Based Regulation," 2001; and "Performance-Based Rates," 2007.

8. In actuality it isn't anywhere near automatic, but it works. Regulators have to first make sure that the savings (9,000 kWh versus 10,000 kWh forecasted) come from the utility's EE efforts, not unusually mild weather or something else. This takes some verification efforts. After this, the calculations and administrative procedures are fairly straightforward. For additional information, see Smith, "Less Demand, Same Great Revenue," 2009; Weston, "Customer-Sited Resources," 2008; and NARUC, "Decoupling for Electric & Gas Utilities," 2007.

9. National Grid, "General Information Pack," 2009: 61.

10. FERC, "Smart Grid Policy," 2009.

11. Ad 2: Progress Energy. http://www.progress-energy.com/aboutus/ads/index.asp

The Energy Services Utility

1. *Power and Energy Magazine*, "Store and Deliver."

2. Darbee. "PG&E's Vision: 2008 and Beyond," 2009.

3. Ibid.

4. Thompson, "A Green Coal Baron," 2008; Schlosser, "How Duke Energy's CEO Got Started," 2009; Thompson, "Meet the Maverick," 2007; and "The Best CEOs in America," 2007.

5. "Coming Soon from a Utility," 2006.

6. Izzo, "Climate Change," 2007.

7. Andres Carvallo, "Austin Energy Plans Its Smart Grid 2.0," 2009.

8. H. Platt, *The Electric City*, p. 22ff.

9. Wasik, "The Merchant of Power: Sam Insull, Thomas Edison, and the Creation of the Modern Metropolis," 2006.

10. The seminal works were Lovins's *Soft Energy Paths* and Roger Sant's 1983 article, "The Coming Market for Energy Services," *Harvard Business Review*, May–June 1980. My recent contribution to the discussion, "Fix Utilities before They Need a Rescue," also appears in the *Harvard Business Review*, July–August 2009.

11. The color rendition index (CRI) is a numerical measurement of the similarity of light to sunlight, which is considered the best light for work tasks and color appreciation. Sunlight has a CRI of 100, halogen bulbs have a CRI over 90, and good compact fluorescent bulbs have a CRI better than 80.

12. As we saw in Chapter 11, National Grid owns no generation and operates entirely in retail choice states. Nonetheless, its chairman, Steve Holliday, recently noted, "I can see a situation where National Grid becomes more of a total provider of energy services to our customers, a holistic manager of energy demand. We will get paid for investing in the right technology at the right time, and providing the right advice to help businesses and homeowners conserve energy, and for all the activity of getting a better handle on the overall supply-and-demand balance." See Burr, "Greenhouse Gauntlet," 2007.

Conclusion

1. Charles Phillips provides a classic overview of the role and function of independent regulatory commissions: *The Regulation of Public Utilities* (Arlington, VA: Public Utilities Reports, Inc.) 1993: 131; and Professor Janice Beecher, director of the Institute of Public Utilities at Michigan State University, explains the modern Commission's duties at length in "Prudent Regulator," 2008.

2. A small selection of recent episodes in which commissioners were criticized or their removal was attempted includes "Analysis: Market Regulation," *Energy Economist*, 2006; RMR, "The Corruption of Power"; and Jamison and others, "Disbanding the Maryland Public Service Commission," 2006.

3. There are many measures that can help professionalize commissions and strike the proper balance between commission stakeholders, such as staggered appointments

and requirements for bipartisan or independent appointments. Where commissioners are elected the more common process of appointment and confirmation should be considered.

4. The costs and challenges of meeting climate change limits will be different for every utility within every segment of the industry, so generalizations and figures applying to entire segments should not be taken out of context. With this caveat, I note that the overall public power sector has the most balanced current carbon footprint due to the fact that it has rights to quite a lot of zero-carbon hydroelectric power. Nationwide, the sector gets 45% of its power from coal, 18% from hydro, 17% from nuclear power, and 16% from natural gas. The cooperative sector is much more carbon-intensive, with 80% of its power generated by coal, 13% from nuclear power, and 7% from gas (as of 2005). American Public Power Association 2009–10 statistical report, p. 24; and NRECA G&T Profiles 2005 from the NRECA Strategic Analysis, February, 2008 (http://www.nreca.org/Coop_bythenos.doc).

5. Rural cooperatives and public power companies have extensive investments in large traditional power plants. These entities do not have much, if any, shareholder equity on their balance sheets. In contrast, their power plants are often financed by the issuance of bonds. The revenues to repay these bonds usually come from selling kilowatt-hours, not from other funding sources. The need to meet bond payment obligations and the inability to draw on shareholder equity if revenues drop below target levels can discourage attempts to lower energy sales.

6. Indeed, a community energy system (CES) is nearly identical to a municipal utility or a cooperative, except that a CES is added into an area where the wires may be owned and operated on an open access basis by a separate Smart Integrator. In this sense, it is the unlikely child of public power and retail choice.

7. "American Electric Power Company, Inc.: Ohio Power Company and the Buckeye project agreements and bond closing documents, volume 1," Lehman Brothers Collection, Harvard Business School Baker Library Historical Collections, 1968.

Appendix A

1. Peter Fox-Penner and Marc Chupka, comment on "Preventing Windfalls for Polluters but Preserving Prices—Waxman–Markey Gets It Right with Its Allocations to Regulated Utilities," Climate Progress, comment posted on May 27, 2009, http://climateprogress.org/2009/05/27/exclusive-report-foxpenner-chupka-waxman-markey-utility-allowances/.

2. See, for example, EIA, 2009(h): v.

3. See, for example, Harrington, Morgenstern, and Nelson, "On the Accuracy of Regulatory," 2000: 297–322.

4. See Neenan and Eom, "Price Elasticity of Demand," 2008; and Fox-Penner, Hledik, and Cajkusic. "Impact of Price Elasticity," 2008, for recent surveys of the literature and discussion.

5. Nadel and Pye, "Appliance and Equipment Efficiency," 1996; and Eldridge et al., "The 2008 State Energy," 2008.

6. EPRI, "Assessment of Achievable Potential," 2009. *The Brattle Group*'s Ahmad Faruqui served as principal investigator in this study.

7. ACEEE et al., "Joint Comments of the American Council," 2009(d).

8. ACEEE, "H.R. 2454 Addresses Climate," 2009(e).

9. MaCabrey, "Soaring electricity," 2009.

10. Siddiqui, Parmenter, and Hurtado, "The Green Grid," 2008.

11. Laitner, Poland-Knight, McKinney, and Ehrhardt-Martinez, "Semiconductor Technologies," 2009.

12. See Faruqui, Sergici, and Sharif, "The Impact of Information Feedback," 2009; and Faruqui, Fox-Penner, and Hledik, "Smart-Grid Strategy," 2009: 32–36, 60.

13. It is also true that onsite DG will need backup power and/or onsite storage, and that either of these may be purchased from utilities. No attempt is made to deny or exclude backup power supplied by utilities as a result of increased DG. Mechanically, added onsite generation is deducted from total electricity used and sold in the EIA reference. By this method, it is automatically the case that energy not generated onsite (i.e., backup energy) is purchased conventionally. Since our interest is only in physical energy sales, the fact that DG customers may incur capacity or other standby charges does not enter the arithmetic in this appendix, but is inherent in the decisions of customers to adopt DG technologies or purchase utility power.

14. EIA's, 2009(k) and Table 8. "Electricity Supply," 2009(g).

Bibliography

2008 Behavior, Energy and Climate Change Conference presentations, November 16–19, 2008, Sacramento, CA.

"A Milestone for Cleaner Coal Technology: Key Equipment Arrives at Duke Energy's IGCC Plant in Indiana." *Globe Investor*, GlobeInvestor.com, June 24, 2009.

AEP. "History of AEP." http://www.aep.com/about/history.

"Achievements: Controlling Human Impact on the Environment." Electricité de France, January 1, 2007. http://sustainable-development.edf.com/accueil-com-fr/edf-developpement-durable/edf-sustainable-development/base-documentaire-en/achievements/controlling-human-impact-on-the-environment/the-plan-aleas-climatiques-extreme-weather-network-plan-107740.html.

"Advanced Metering and Demand Responsive Infrastructure: A Summary of the Pier/CEC Reference Design, Related Research and Key Findings." Draft, Prepared for California Energy Commission, EnerNex Corporation, June 1, 2005.

Alexander, Lamar. "Blueprint for 100 New Nuclear Power Plants in 20 Years." Prepublication copy, U.S. Senate, July 13, 2009 http://alexander.senate.gov/public/_pdfs/blueprint.pdf.

American Council on Renewable Energy (ACORE). "The Outlook on Renewable Energy in America." January 2007.

American Council for an Energy-Efficient Economy (ACEEE). "H.R. 2454 Addresses Climate Change through a Wide Variety of Energy Efficiency Measures." American

Council for Energy Efficiency, June 1, 2009(e). http://aceee.org/energy/national/ HR2454_Estimate06-01.pdf.

————, et al. "Joint Comments of the American Council for an Energy-Efficient Economy, the Alliance to Save Energy, the Natural Resources Defense Council and Energy Center of Wisconsin on the January 2009 Report; 'Assessment of Achievable Potential from Energy Efficiency and Demand Response Programs in the U.S.' issued by EPRI." 2009(d).

————. "Quantifying the Effect of Market Failures and in the End-Use of Energy." February 2007.

————. "Savings Estimates for ACESA Rules' Committee Version (Waxman–Markey), Discussion Draft Analysis Chart, June 24, 2009(a).

————. "Success with Energy Efficiency Resource Standards." January 2009(c).

————. "U.S. DOE Heats Up New Energy Standards for Home Furnaces." ACEEE News Release, April 23, 2009(b).

American Electric Power (AEP). "Interstate Transmission Vision for Wind Integration." http://www.aep.com/about/i765project/docs/WindTransmissionVisionWhitePaper .pdf.

"American's Energy Future: Technology and Transformation." National Research Council of The National Academies, 2009. http://www.nap.edu.

American Public Power Association (APPA). "2009–2010 APPA Annual Director & Statistical Report." 2009: 24.

————. "APPA Comments to NERC on Reliability Impacts of Climate Change Initiatives." July 16, 2008. http://www.appanet.org/files/PDFs/APPACommentsNERC ReliabilityImpactsClimateChange71608.pdf.

————. "Restructuring at the Crossroads: FERC Electric Policy Reconsidered." December 2004.

Amin, S. Massoud. "For the Good of the Grid." *IEEE power & energy magazine*, November/December 2008: 53–54.

"Analysis: Market Regulation." *Energy Economist*, November 1, 2006.

Apt, Jay. "Competition Has Not Lowered U.S. Industrial Electricity Prices." *Electricity Journal*, March 2005: 52–56.

Apt, Jay, et al., "Electrical Blackouts: A Systemic Problem." *Issues in Science and Technology*, June 22, 2004.

Arnold, Jon, and Larry Cochrane. "Future Opportunities for the Energy Ecosystem." *Power Engineering*, July 2009.

Arocena, Pablo. "Cost and Quality Gains from Diversification and Vertical Integration in the Electricity Industry: A DEA Approach." *Energy Economics*, January 2008: 39–58.

"AWEA Small Wind Turbine Global Market Study." American Wind Energy Association (AWEA), 2009.

Azagury, Jack, et al. "Building the Next Generation Utility." *Public Utilities Fortnightly*, January 2009.

"Baltimore Gas and Electric Company Unveils Plans for One of the Most Advanced Smart Grid Initiatives in the Nation." *Business Wire*, July 13, 2009.

Bandivadekar, A., K. Bodek, L. Cheah, C. Evans, T. Groode, J. Heywood, E. Kasseris, M. Kromer, and M. Weiss. *On the Road in 2035: Reducing Transportation's Petroleum Consumption and GHG Emissions.* MIT Laboratory for Energy and the Environment. Cambridge, Massachusetts: 2008. http://web.mit.edu/sloan-auto-lab/research/beforeh2/otr2035/.

Barmack, Matthew, Edward Kahn and Susan Tierney. "A Cost–Benefit Assessment of Wholesale Electricity Restructuring and Competition in New England." Analysis Group, 2006.

Basheda, G., M. Chupka, P. Fox-Penner, H. Pfeifenberger, and A. Schumacher. "Why Are Electricity Prices Increasing? An Industry Perspective." Prepared for The Edison Foundation, by *The Brattle Group*, June 2006.

Baxter, Richard. "Energy Storage: An Expanding Role as a Distributed Resource." *Cogeneration and On-Site Power Production*, May–June 2003: 69–72.

Beattie, Jeff. "Thanks, But No Thanks: Southern, AEP Exit FutureGen." *Energy Daily*, June 26, 2009.

———. "Only Six New Nukes to Break Ground Immediately—NRC." *Energy Daily*, June 3, 2009.

———. "Utilities Unprepared for New Reactor Costs—Moody's." *Energy Daily*, August 14, 2009.

Beck, Michael J., and William Klun. "IOUs under Pressure." *Public Utilities Fortnightly*, June 2009: 37–41.

Beder, Sharon. "The Electricity Deregulation Con Game." *PR Watch*, Third Quarter 2003, 10, (3).

Beecher, Janice A. "Avoided Cost: An Essential Concept for Integrated Resource Planning." Center for Urban Policy and the Environment, Indianapolis, Indiana University–Purdue University, 1995.

———. "The Prudent Regulator: Politics, Independence, Ethics, and the Public Interest." *Energy Law Journal*, July 1, 2008: 577–614.

Behr, Peter. "Probe of California Energy Crisis Facing Hurdles." *Washington Post*, January 11, 2003: E01.

Benton, James C. "Senator Certain of Savings: Others Not So Sure Electric-Deregulation Bills Benefit Customers, Schools." *Akron Beacon-Journal*, March 29, 1999.

Bertani, Ruggero. "Long-Term Projections of Geothermal-Electric Development in the World." Enel Green Power, presented at GeoTHERM Conference, Offenburg 2009.

Bezdek, Roger H. "Green Collar Jobs in the U.S. and Colorado, Economic Drivers for the 21st Century." American Solar Energy Society, January 2009.

————. "Renewable Energy and Energy Efficiency: Economic Drivers for the 21st Century." American Solar Energy Society, 2007.

Biello, David. "How Fast Can Carbon Capture and Storage Fix Climate Change?" *Scientific American*, April 10, 2009.

Blumsack, Seth A., Jay Apt, and Lester B. Lave. "Lessons from the Failure of U.S. Electricity Restructuring." *Electricity Journal*, March 2006: 51–32.

Blumstein, Carl, L. S. Friedman, and R. J. Green. "The History of Electricity Restructuring in California." CSEM WP 103, UCEI, August 2002.

Boedecker, Erin, John Cymbalsky, and Steven Wade. "Modeling Distributed Electricity Generation in the NEMS Buildings Model." EIA Forecast, Energy Information Administration, U.S. Department of Energy, July 30, 2002. http://www.eia.doe.gov/oiaf/analysispaper/electricity_generation.html.

Bohn, Roger E., Alvin K. Klevorick, and Charles G. Stalon. "Second Report on Market Issues in the California Power Exchange Energy Markets." Prepared for the Federal Energy Regulatory Commission, March 19, 1999.

"Bottling Electricity: Storage as a Strategic Tool for Managing Variability and Capacity Concerns in the Modern Grid." Electricity Advisory Committee, December 2008.

Borenstein, Severin, and James Bushnell. "Electricity Restructuring: Deregulation or Reregulation?" *Regulation*, 23 (2), 2000.

Boyle, Matthew. "A smarter, greener grid." *Fortune*, May 7, 2008.

Bradford, Peter A. "The Myth(s) of the Nuclear Renaissance." Environmental and Energy Study Institute (EESI) Briefing, Washington, DC, May 21, 2009.

Bramble, Barbara J. "Roundtable for Sustainable Biofuels." Presented at BIOMASS 2009: Fueling Our Future, March 17–18, 2009.

Brennan, Tim. "Questioning the Conventional 'Wisdom,'" *Regulation*, Fall 2001.

Brockway, Nancy. "Advanced Metering Infrastructure: What Regulators Need to Know about Its Value to Residential Customers." National Regulatory Research Institute, February 13, 2008.

"Brodsky Introduces Reform Legislation to Reduce Electric Rates and Re-regulate Energy Industry." *Yonkers Tribune*, November 8, 2007. http://yonkerstribune.typepad.com/yonkers_tribune/2007/11/brodsky-introdu.html.

Brown, Garry A. Testimony before the U.S. Senate, Subcommittee on Energy, Committee on Energy and Natural Resources on "Net Metering, Interconnection Standards, and Distributed Generation." May 7, 2009.

Brown, Marilyn A. "Market Failures and Barriers as a Basis for Clean Energy Policies," *Energy Policy*, November 2001: 1197–1207.

Brown, Matthew, H. "California's Power Crisis: What Happened? What Can We Learn?" National Conference of State Legislatures, March 2001.

Brubaker, Harold. "After the Bubble." *Philadelphia Inquirer*, May 3, 2009.

Building Codes Assistance Project. "Commercial Building Energy Codes—Usability and Compliance Methods." October 2008.

————. "Residential Building Energy Codes—Enforcement and Compliance Study." October 2008.

Burr, Michael T. "Conservation Compact: Utilities Test New Models to Encourage Investment in Efficiency and Conservation." *Public Utilities Fortnightly*, June 2008: 39–74.

————. "Greenhouse Gauntlet." *Public Utilities Fortnightly*, June 2007: 41–51.

Bushnell, James, Alvin K. Klevorick, and Robert Wilmouth. "Third Report on Market Issues in the California Power Exchange Energy Markets." Prepared for the Federal Energy Regulatory Committee, March 6, 2000.

Bushnell, James B., and Frank A. Wolak. "Regulation and the Leverage of Local Market Power in the California Electricity Market." UC Berkeley, Competition Policy Center, July 1999.

"CEE Cost Effectiveness Tools." Updated June 28, 2007. http://222.ethree.com/cpuc_avoidedcosts.html.

California Energy Commission. "Renewable Energy Transmission Initiative—Phase IB, Final Report." January 2009. http://www.energy.ca.gov/2008publications/RETI-1000-2008-003/RETI-1000-2008-003-F.PDF.

California ISO (CAISO). "Alerts, Warnings, and Emergencies." April 2004. http://www.caiso.com/awe/AlertsWarnings-WhitePaper.pdf.

————. "California ISO Urges Continued Conservation: Friday is Day 32 of Stage Three Electrical Emergency." February 15, 2001. http://www.caiso.com/docs/2001/02/15/200102151719381517.pdf.

California Public Utilities Commission. "Interim Opinion on E3 Avoided Cost Methodology." Rulemaking 04-04-225, April 22, 2004.

————. California Solar Initiative, Program Handbook. July 2009: 19, Sec. 2.2.3.

California State Senate. Steve Peace Bio. 2002. http://www.sen.ca.gov/ftp/sen/SENATOR/_ARCHIVE_2002/DEPARTING/PEACE/BIO.htm.

Cappers, Peter, Charles Goldman, and David Kathan. "Demand Response in U.S. Electricity Markets: Empirical Evidence." Lawrence Berkeley National Laboratory, LBNL-2124E, *Energy*, July 12, 2009.

"Carbon Capture and Geological Storage in Emerging Countries: Financing the EU–China Near Zero Emissions Coal Plant Project." European Commission, *eGov Monitor*, June 26, 2009. http://www.egovmonitor.com/node/25910.

Carey, John. "A Smarter Electric Grid." *BusinessWeek*, January 11, 2008.

Carlisle, N, J. Elling, and T. Penny. "A Renewable Energy Community: Key Elements." NREL/TP-540-42774, NREL, January 2008: 1–2.

Carvallo, Andres. "Austin Energy Plans Its Smart Grid 2.0." CIO Master and Smart Grid Master Blog, April 18, 2009. http://www.coimaster.com/2009/04/austin-energy-plans-its-smart-grid-20.html.

Casten, Sean. "CHP: One of the answers (but not the question)." Presented to Efficient Enterprises: Power American Industry, June 23, 2009.

Cavanagh, Ralph. "Energy Efficiency in Buildings and Equipment: Remedies for Pervasive Market Failures." Prepared for the National Commission on Energy Policy, December 1, 2004.

Celebi, Metin, and Frank Graves. "CO_2 Price Volatility: Consequences and Cures." *The Brattle Group*, January 2009.

Centolella, Paul, and Andrew Ott. "The Integration of Price Responsive Demand into PJM Wholesale Power Markets and System Operations." March 9, 2009: 4. http://www.hks.harvard.edu/hepg/Papers/2009/Centolella%20%20Ott%20PJM%20PRD%2003092009.pdf.

"Challenges For Sustainable Second Generation Biofuels." CSBP, presented at Biomass 2009, Washington, DC, March 17, 2009.

Chandley, John D., Carl R. Danner, Christopher E. Groves, et al. "California's Electricity Markets: Structure, Crisis, and Needed Reforms." LECG, LLC., January 17, 2003.

Chassin, David P., and Lynne Kiesling. "Decentralized Coordination through Digital Technology, Dynamic Pricing, and Customer-Drive Control: The GridWise Testbed Demonstration Project." *Electricity Journal*, October 2008: 51–59.

Cheng, Roger. "Cisco Looking to Tap Growing Smart-Grid Market." *Wall Street Journal*, May 18, 2009.

Chupka, Marc. "Uncertainty, Volatility and Risk: Cap and Trade Economics." *The Brattle Group*, presented to The Aspen Institute Energy Policy Forum, July 8, 2008.

Chupka, Marc, et al. "Transforming America's Power Industry: The Investment Challenge 2010–2030." *The Brattle Group*, Prepared for The Edison Foundation, November 2008.

Cicchetti, Charles J., and Colin M. Long, with Kristina M Sepetys. *Restructuring Electricity Markets: A World Perspective Post California and Enron*. Visions Communications, 2003.

Claeys, Richard. "Changing Course: Latest RKS Survey of State Utility Regulators Documents Retreat from Deregulation." *Business Wire*, September 19, 2007. http://findarticles.com/p/articles/mi_m0EIN/is_2007_Sept_20/ai_n27380541.

Clark, Wilson. *Energy for Survival*. (Garden City: Anchor Books) 1974.

Claussen, Eileen. "Energy Efficiency, Climate Change and Our Nation's Future." Presented at Energy Efficiency Forum, Washington, DC, June 16, 2004.

Cleetus R., S. Clemmer, and D. Friedman. "Climate 2030: A National Blueprint for a Clean Energy Economy." Union of Concerned Scientists (UCS), May 2009.

"Climate Change 2007: Synthesis Report, Summary for Policymakers." IPCC Working Group, November 2007.

Coase, R. H. *The Firm, the Market, and the Law*. (Chicago: University of Chicago Press) 1988.

Cohen, Armond, Mike Fowler, and Kurt Waltzer. "'NowGen': Getting Real about Coal Carbon Capture and Sequestration." *Electricity Journal*, May 2009: 25–41.

Cohen, Linda R., and Roger G. Noll. "Privatizing Public Research." *Scientific American*, September 1994.

"Combined Heat and Power Market Potential for Opportunity Fuels." Resource Dynamics Corporation, August 2004.

"Coming Soon from a Utility Near You: More Power to the People." Knowledge@W.P. Carey, October 11, 2006. http://knowledge.wpcarey.asu.edu/article.cfm?articleid=1310.

Committee Workshop on Combined Heat and Power, Notice of Committee Workshop, Docket No. 09-IEP-1H, California Energy Commission, July 23, 2009.

"Competitive Electricity Markets Drive Renewables, Demand Response, Conservation, Efficiency and Innovation." Compete, May 20, 2009. http://www.compete coalition.com/resources/competitive-electricity-markets-drive-renewables-demand-response-conservation-efficiency-a.

Condon, Stephanie. "Lack of Standards Could Stymie Smart Grid." CNET News, March 3, 2009. http://news.cnet.com/8301-13578_3-10187292-38.html.

Congressional Budget Office. "Nuclear Power's Role in Generating Electricity." A CBO Study, Washington, DC, May 2008.

"Cooling, Heating, and Power for Industry: A Market Assessment." Resource Dynamics Corporation, August 2003.

Cooper, Mark. "The Economics of Nuclear Reactors: Renaissance or Relapse?" Institute for Energy and the Environment, Vermont Law School, June 2009.

Creyts, Jon, Anton Derkach, Scott Nyquist, Ken Ostrowski, and Jack Stephenson. "Reducing U.S. Greenhouse Emissions: How Much at What Cost?" U.S. Greenhouse Gas Abatement Mapping Initiative, Executive Report. McKinsey & Company and The Conference Board, 2007. http://www.mckinsey.com/clientservice/ccsi/pdf/US_ghg_final_report.pdf.

Cudahy, Richard D. "The Coming Demise of Deregulation." *Yale Journal on Regulation*, Issue 10.1, 1993.

"Cutting CO_2 emissions from existing coal plants." MIT News, June 19, 2009. http://web.mit.edu/newsoffice/2009/coal-0619.html.

Dao, James. "The End of the Last Great Monopoly." *New York Times*, August 4, 1996.

Darbee, Peter A. "PG&E's Vision: 2008 and Beyond." Pacific Gas and Electric Corporation, February 26, 2009.

"Dark Days Ahead." *Economist*. August 6, 2009.

Davis, Bob, Damian Paletta, and Rebecca Smith. "Unraveling Reagan: Amid Turmoil, U.S. Turns Away from Decades of Deregulation." *Wall Street Journal*, July 25, 2008.

"Deciding on 'Smart' Meters: The Technology Implications of Section 1252 of the Energy Policy Act of 2005." Prepared by: Plexus Research, Inc., for Edison Electric Institute, September 2006.

deLaski, Andrew. Testimony before the U.S. House of Representatives, Energy and Commerce Committee; Hearing on: American Clean Energy and Security Act of 2009,
April 24, 2009.

Dennis, Keith. "The Compatibility of Economic Theory and Proactive Energy Efficiency
Policy." *Electricity Journal*, Aug./Sep. 2006: 58–72

Dernbach, John C. "Energy Efficiency and Conservation as Ethical Responsibilities: Suggestions for IPCC Working Group III." January 27, 2008. http://papers.ssrn.com/
sol3/papers.cfm?abstract_id=1089423.

Deutch, John M., et al. "Update of the MIT 2003 Future of Nuclear Power." MITEI, Massachusetts Institute of Technology, 2009.

"DOE Providing $408 Million for Two CO_2 Capture Projects." *Energy Daily*, July 2,
2009: 3.

"DOE: Residential Consumers in All States to Benefit from Deregulation." *Utility Spot-
light*, 1999.

Douglas, Paul, et al. "33% Renewables Portfolio Standard: Implementation Analysis Preliminary Results." California Public Utilities Commission, June 2009.

"Duke Energy's Jim Rogers Get Coveted 'CEO of the Year.'" Media Relations and Event
Management, November 30, 2007.

Duke, Rick, and Dan Lashof. "The New Energy Economy: Putting America on the Path
to Solving Global Warming." NRDC Issue Paper, May 2008.

Duncan, Ian. "Carbon Sequestration Risks, Opportunities, and Learning from the CO_2–
EOR Industry." To the Committee on Energy and Commerce, U.S. House of Representatives, March 10, 2009.

"Economic Security: Will the Third Oil Shock Be the Charm?" Clean Fuels Development
Coalition, September 2009.

Edison Electric Institute (EEI). "Transmission Projects: Supporting Renewable Resources." February 2009.

Ehrhardt-Martinez, Karen, and John A. "Skip" Laitner. "The Size of the U.S. Energy Efficiency Market: Generating a More Complete Picture." American Council for an
Energy-Efficient Economy, E083, May 2008.

Eldridge, Maggie, et al., "The 2008 State Energy Efficiency Scorecard." American Council
for an Energy-Efficient Economy, October 2008.

Electric Power Research Institute (EPRI). "Advanced Coal Power Systems with CO_2 Capture: EPRI's Coal Fleet for Tomorrow Vision." Interim Report, Palo Alto, CA: September 2008(a): 5, 1–5.

———. "Assessment of Achievable Potential from Energy Efficiency and Demand Response Programs in the U.S. (2010–2030)." Technical Report, Palo Alto, CA: January
2009(a).

———. "Environmental Assessment of Plug-In Hybrid Electric Vehicles Volume 1:
Nationwide Greenhouse Gas Emissions." Technical Report, Palo Alto, CA: July
2007(a).

———. "Environmental Assessment of Plug-In Hybrid Electric Vehicles Volume 2: United States Air Quality Analysis Based on AEO-2006 Assumptions for 2030." Technical Report, Palo Alto, CA: July 2007(b).

———. "Interim Smart Grid Roadmap." Draft Report, Electric Power Research Institute (EPRI), March 24, 2009(b).

———. "Program on Technology Innovation: Integrated Technology Options." Technical Update, Palo Alto, CA: November 2008.

Electricity from Renewable Resources: Status, Prospects, and Impediments. (Washington, DC: National Academies Press) 2009.

"Empowering Consumers Through a Modern Electric Grid." Report of the Illinois Smart Grid Initiative. April 2009: 20–21.

Energy Information Administration (EIA). "An Updated Annual Energy Outlook 2009 Reference Case Reflecting Provisions of the American Recovery and Reinvestment Act and Recent Changes in the Economic Outlook." *The Annual Energy Outlook 2009.* SR-OIAF/2009-03 (2009). Washington, DC, April 2009(g). (Tables 2, 8, 45.)

———. "Annual Energy Outlook 2009 Early Release Overview." DOE/EIA-0484 (2009). Washington, DC, January 2009(c). http://www.eia.doe.gov/oiaf/ieo/.

———. "Nuclear Energy." *Annual Energy Review 2008.* Section 9, DOE/EIA-0384(2008), Washington, DC, June 2009(j).

———. "Appendix A: History of the U.S. Electric Power Industry, 1882–1991." *The Changing Structure of the Electric Power Industry 2000: An Update.* Washington, DC, January 2009(a). http://www.eia.doe.gov/cneaf/electricity/chg_stru_update/appa .html.

———. "Assumptions to the Annual Energy Outlook 2009." DOE/EIA-0554(2009), Washington, DC, March 2009(f). http://www.eia.doe.gov/oiaf/aeo/assumption/ pdf/0554(2009).pdf.

———. "Biomass Milestones." http://www.eia.doe.gov/cneaf/solar.renewables/ renewable.energy.annual/backgrnd/chap6.htm.

———. "Carbon Capture & Storage." Presented at NARUC Winter Meeting, Washington, DC, February 17, 2009(e).

———. "Energy Market and Economic Impacts of H.R. 2454, the American Clean Energy and Security Act of 2009." Errata, SR/OIAF/2009-5, U.S. Department of Energy, August 2009(k).

———. "Energy Market and Economic Impacts of S. 2191, the Lieberman–Warner Climate Security Act of 2007." SR/OIAF/2008-01, Washington, DC, April 2008(a).

———. "Existing Capacity by Producer Type." Energy Information Administration, U.S. Department of Energy, January 21, 2009(d): http://www.eia.doe.gov/cneaf/ electricity/epa/epat2p3.html.

———. "History of Energy in the United States: 1635–2000 Total Energy." Washington, DC, September 2003. http://www.eia.doe.gov/emeu/aer/eh/frame.html.

————. "Impacts of a 25-Percent Renewable Electricity Standard as Proposed in the American Clean Energy and Security Act Discussion Draft." SR/OIAF/2009-04, Washington, DC, April 2009(h).

————. "New Commercial Reactor Designs." November 2006. http://www.eia.doe.gov/cneaf/nuclear/page/analysis/nucenviss2.html.

————. Status of Electricity Restructuring by State, May 2009(i). http://www.eia.doe.gov/cneaf/electricity/page/restructuring/restructure_elect.html.

————. Table A2: "Energy Consumption by Sector and Source." *The Annual Energy Outlook 2009–Early Release*, Washington, DC, December 17, 2008(d).

————. Table 2.2: "Existing Capacity by Energy Source, 2007." *Electric Power Annual 2007*, Washington, DC, January 26, 2007.

————. Table 5: "U.S. Average Monthly Bill by Sector, Census Division and State, 2007." *Electric Sales, Revenue, and Price Report*. Washington, DC, January 2009(b).

————. Table 127: "Imported Liquids by Source." Supplemental Table to the Annual Energy Outlook 2010, Washington, DC, December 14, 2009(l). http://www.eia.doe.gov/oiaf/aeo/supplement/supref.html.

————. "The Electric Utility Industry Restructuring Act." See Provisions of AB 1890, Status of the California Electricity Situation, 1998. http://www.eia.doe.gov/cneaf/electricity/california/assemblybill.html.

————. "What Are Greenhouse Gases and How Much Are Emitted by the United States?" *Energy in Brief—What Everyone Should Know about Energy*. Washington, DC, July 10, 2008(c). http://tonto.eia.doe.gov/energy_in_brief/greenhouse_gas.cfm.

————. "World Energy Demand and Economic Outlook." *International Energy Outlook 2008*. DOE/EIA-0383 (2008), Washington, DC, June 2008(b).

"Energy Security Wars: Grids vs. Hackers." The DOD Energy Blog, March 10, 2009. http://dodenergy.blogspot.com/2009/03/energy-security-wars-grids-vs-hackers.html.

Fairley, Peter. "Germany's Green-Energy Gap." IEEE *Spectrum*, July 2009. http://spectrum.ieee.org/energy/policy/germanys-green-energy-gap/0.

"False Consciousness: Everybody Loved ENRON." *Brandweek*, March 3, 2006.

Faruqui, Ahmad. "Breaking Out of the Bubble." *Public Utilities Fortnightly*, March 2007: 46–51.

————. "Creating Value through Demand Response." Presented at DOE's Smart Grid E-Forum, October 23, 2008.

————. "Inclining Toward Efficiency." *Public Utilities Fortnightly*, August 2008: 22–27.

————. "Will the Smart Grid Promote Wise Energy Choices?" Presented at Illinois Smart Grid Initiative, Chicago, IL, August 5, 2008.

Faruqui, Ahmad, Hung-po Chao, Vic Niemeyer, Jeremy Platt, and Karl Stahlkopf. "Analyzing California's Power Crisis." *Energy Journal*, 22 (4), 2001: 29–51.

Faruqui, Ahmad, and Lisa Wood. "Quantifying the Benefits of Dynamic Pricing In the Mass Market." *The Brattle Group*, January 2008.

Faruqui, Ahmad, and Ryan Hledik. "Transition to Dynamic Pricing." *Public Utilities Fortnightly*, March 2009: 27–33.

Faruqui, Ahmad. Ryan Hledik, and Clay Davis. "Sizing Up the Smart Grid." Presentation from *The Brattle Group*, March 17, 2009.

Faruqui, Ahmad, Peter Fox-Penner, and Ryan Hledik, "Smart-Grid Strategy: Quantifying Benefits." *Public Utilities Fortnightly*, July 2009: 32–36, 60.

Faruqui, Ahmad, Ryan Hledik, and John Tsoukalis. "The Power of Dynamic Pricing." *Electricity Journal*, April 2009: 42–56.

Faruqui, Ahmad, Ryan Hledik, Sam Newell, and Johannes Pfeifenberger. "How Dynamic Pricing Can Save $35 Billion in Electricity Costs." *The Brattle Group* Discussion Paper, May 16, 2007.

Faruqui, Ahmad, Sanem Sergici, and Ahmad Sharif, "The Impact of Information Feedback on Energy Consumption—A Survey of the Experimental Evidence." *The Brattle Group* 2009.

Faruqui, Ahmad, Sanem Sergici, and Lisa Wood, "Moving Toward Utility-Scale Deployment of Dynamic Pricing in Mass Markets." IEE Whitepaper, June 2009.

Federal Energy Regulatory Commission (FERC). "Electric Market Overview: Renewable Portfolio Standards." July 8, 2009. http://www.ferc.gov/market-oversight/mkt-electric/overview/elec-ovr-rps.pdf.

———. "Final Report on Price Manipulation in Western Markets." Docket No. PA02-2-000, March 2003.

———. "Order on Cost Filings." Docket No. EL00-95-000, et al. January 26, 2006.

———. "Smart Grid Policy." Docket No. PL09-04-000, March 19, 2009.

Federal Ministry for the Environment, Nature Conservation and Nuclear Safety (BMU). "Renewable Energy: Employment Effects." Berlin, Germany, June 2006.

Fehrenbacher, Katie. "How to Hammer Out Smart Grid Standards in 30 Days or Less, Or Your Money Back." Earth2tech.com, May 7, 2009. http://earth2tech.com/2009/05/07/how-to-hammer-out-smart-grid-standards-in-30-days-or-less-or-your-money-back/.

Finon, Dominique. "Investment Risk Allocation in Decentralised Electricity Markets: The Need of Long-Term Contracts and Vertical Integration." *OPEC Review*, June 2008: 150–183.

Fisher, Jolanka V., and Timothy P. Duane. "Trends in Electricity Consumption, Peak Demand and Generation Capacity in California and the Western Grid." Program on Workable Energy Regulation (POWER), PWP-85, September 2001.

Fleischauer, Eric. "Heat Wave Shutdown at Browns Ferry Stirs Nuclear Debate." Climate Ark: Climate Change and Global Warming Portal, September 2, 2007. http://www.climateark.org/shared/reader/welcome.aspx?linkid=83238&keybold=climate%20blogs.

Fleischmann, Daniel J. "An Assessment of Geothermal Resource Development Needs in the Western United States." January 2007.

Flint, Alex. Statement before the Senate Republican Conference, Nuclear Energy Institute, June 8, 2009.

Forsyth, Trudy. "Small (Distributed) Wind Technology." National Renewable Energy Laboratory, June 2008.

Fox-Penner, Peter. "A Welcome Truce in the Electricity Wars." *Public Utilities Fortnightly*, 2005: 48–51.

———. "Allowing for Regulation in Forecasting Load and Financial Performance." *Public Utilities Fortnightly*, January 7, 1988.

———. "Electric Power Deregulation: Blessings and Blemishes, A Non-Technical Review of the Issues Associated with Competition in Today's Electric Power Industry." Prepared for the National Council on Competition and the Electric Industry, March 14, 2000.

———. *Electric Power Transmission and Wheeling: A Technical Primer.* (Edison Electric Institute) 1990.

———. *Electric Restructuring: A Guide to the Competitive Era.* (Vienna, VA: Public Utility Reports) 1997.

———. "Emerging Smart Grid Technologies: The Future of U.S. Power Distribution." Presented to the 2degrees Intelligent Grids Network Webinar, April 22, 2009.

———. "Fix Utilities before They Need a Rescue." *Harvard Business Review*, July–August 2009: 18–19.

———. Prefiled Rebuttal Testimony before the Public Utilities Commission of Nevada, Nevada Power Company's 2001 Deferred Energy Filing, PUCN Docket No. 01-11029, p. 77.

———. Testimony, Federal Energy Regulatory Commission, Docket No. EL0-95-000, et al., regarding the Investigation of Practices of the CAISO and CALPX.

———. Prepared Rebuttal Testimony, Federal Energy Regulatory Commission, Docket No. EL0-95-000, et al., (CA-349) regarding Market Fundamentals, March 20, 2003.

———. Testimony for Cheyenne Light, Fuel and Power Company. Docket No. 20003-EP-01-59, March 2001.

———. "The Impact of the Public Utility Regulatory Policies Act of 1978 (PURPA): Retrospect and Prospect." Draft, U.S. Department of Energy, January 9, 1996.

———. "Transforming America's Power Industry: The Investment Challenge." Presentation to The Edison Foundation, Phoenix, AZ, November 10, 2008.

———. "U.S. Transmission Investment: Policies and Prospects." *The Brattle Group*, April 28, 2009.

Fox-Penner, Peter, and Marc Chupka. "Preventing Windfalls for Polluters but Preserving Prices—Waxman–Markey Gets It Right with Its Allocations to Regulated Utilities." Climate Progress Blog, Posted May 27, 2009. http://climateprogress.org/2009/05/27/exclusive-report-foxpenner-chupka-waxman-markey-utility-allowances/.

Fox-Penner, Peter, Dean M. Murhpy, Mariko Geronimo, and Matthew McCaffree. "Promoting Use of Plug-In Electric Vehicles through Utility Industry Acquisition and Leasing of Batteries." Chapter 13 of *Plug-In Electric Vehicles: What Role of Washington?*, Washington, DC: The Brookings Institution, 2009.

Fox-Penner, Peter and Johannes Pfeifenberger. Affidavit, Federal Energy Regulatory Commission, Docket No. EC05-43-000, April 11, 2005.

Fox-Penner, Peter, Ryan Hledik, and Igor Cajkusic. "The Impact of Price Elasticity on Electric Industry Investments—Preliminary Findings." Prepared for The Edison Foundation, *The Brattle Group*, November 10, 2008.

Fraquelli, Giovanni, Massimiliano Placenza, and Davide Vannoni. "Cost Savings from Generation and Distribution with an Application to Italian Electric Utilities." *Journal of Regulatory Economics*, November 2005: 289–308.

Fridleifsson, I. B., et al. "The Possible Role and Contribution of Geothermal Energy to the Mitigation of Climate Change." IPCC Geothermal, February 11, 2008.

Friedman, David. "Statement of the Union of Concerned Scientists before the House Committee on Energy and Commerce Subcommittee on Energy and Environment." April 24, 2009.

Friedman, Lisa. "China: A Sea Change in the Nation's Attitude toward Carbon Capture." *Climate Wire*, June 22, 2009. http://www.eenews.net/climatewire/print/2009/06/22/1.

FutureGen Alliance. "FutureGen Technology." 2009. http://www.FutureGenAlliance.org.

Gallagher, Patrick D., Ph.D., Testimony before the Committee on Energy and Natural Resources, U.S. Senate, March 3, 2009.

Galvin, Robert and Kurt Yeager. *Perfect Power*. (McGraw-Hill) 2009.

Gartner, Inc. "Gartner Says More Than 1 Billion PCs In Use Worldwide and Headed to 2 Billion Units by 2014." Gartner, Inc., June 23, 2008. http://www.gartner.com/it/page.jsp?id=703807.

Gellatly, Mike. "Nuclear Industry Needs 10K Workers in 10 Years." *Aiken Standard*, June 12, 2009. http://www.aikenstandard.com/Local/0612Study.

"Generators and ISO New England Worked Well Together during January Freeze, Says a Preliminary Analysis." *Foster Natural Gas Report*, April 1, 2004: 4.

Gillingham, Kenneth, Richard G. Newell and Karen Palmer. "Energy Efficiency Economics and Policy." A Discussion Paper, Resources for the Future, RFF DP 09-13, April 2009.

Goering, Laurie. "Nuclear Plants Being Revived Worldwide." *Chicago Tribune*, March 11, 2009.

Gold, Russell. "Exxon Shale-Gas Find Looks Big." *Wall Street Journal*, July 10, 2009.

Goldstein, Larry, et al. "Gas-Fired Distributed Energy Resources Technology Characterizations." NREL/TP-620-34783, NREL, November 2003.

Golove, William H., and Joseph H. Eto. "Market Barriers to Energy Efficiency: A Critical Reappraisal of the Rationale for Public Policies to Promote Energy Efficiency." LBL-38059, Lawrence Berkeley National Laboratory, March 1996.

Goodell, Jeff. *Big Coal: The Dirty Secret behind America's Energy Future.* (New York: Houghton Mifflin Harcourt) 2007: 101.

Goodin, Dan. "Buggy 'Smart Meters' Open Door to Power-Grid Botnet." *Register*, June 12, 2009. http://www.theregister.co.uk/2009/0612/smart_grid_securitiy_risks/print .html.

Google. "Clean Energy 2030." October 1, 2008. http://knoll.google.com/k/Jeffery-green blatt/clean-energy-2030/15x31uzlqeo5n/1.

Gorman, Siobhan. "Electricity Grid in U.S. Penetrated by Spies." *Wall Street Journal*, April 8, 2009.

Greenberg, Andy. "The Smart Grid vs. Grandma." Forbes.com, May 15, 2009. http:// www.forbes.com/2009/05/15/smart-grid-energy-technology-internet-infrastructure-energy.html.

Greene, Nathanael, and Roel Hammerschlag. "Small and Clean Is Beautiful: Exploring the Emissions of Distributed Generation and Pollution Prevention Policies." *Electricity Journal*, June 2000: 50–60.

Greer, Monica L. "A Test of Vertical Economies for Non-vertically Integrated Firms: The Case of Rural Electric Cooperatives." *Energy Economics*, May 2008: 679–687.

Gronewold, Nathanial. "N.Y.'s Pioneering Effort to Store Power Plant Carbon Faces Legal Gauntlet." *Climate Wire*, April 22, 2009. http://www.eenews.net/cw/2009/04/22.

———. "IPCC Chief Raps G-8, Calls for Global Greenhouse Gas Emission Cuts After 2015." *New York Times*, July 21, 2009.

Gunther, Erich W., et al. "Smart Grid Standards Assessment and Recommendations for Adoption and Development." Prepared for the California Energy Commission, EnerNex Corporation, February 2009.

Gyuk, Imre. "Energy Storage Applications for the Electric Grid." U.S. Department of Energy, October 23, 2008.

Hadley, S. W., J. W. VanDyke, and T. K. Stovall, "Distributed Generation: Benefit Values in Hard Numbers." *Public Utilities Fortnightly*, April 2005: 46–53.

Hadley, Stanton W., and Alexandra Tsvetkova, "Potential Impacts of Plug-in Hybrid Electric Vehicles on Regional Power Generation." Oak Ridge National Laboratory. Oak Ridge, TN; For the U.S. Department of Energy. ORNL/TM-2007/150, January 2008.

Hall, Bronwyn. "The Financing of Research and Development." Working Paper 8773, National Bureau of Economic Research, February 2002. http://www.nber.org/ papers/w8773.

Hamm, Gregory, and Adam Borison. "The Rush to Coal: Is the Analysis Complete?" *Electricity Journal*, Jan./Feb. 2008: 31–37.

Hammerstrom, D., et al. "Pacific Northwest GridWiseTM Testbed Demonstration Projects, Part I: The Olympic Peninsula Project 2007." PNNL-17167, Pacific Northwest National Laboratory, October 2007. http://www.gridwise.pnl.gov/docs/op_project _final_report_pnnl17167.pdf.

Hand, Aaron. "PV Materials to See Exploding Growth." PVSociety.com, May 7, 2009. http://www.pvsociety.com/article/231838-PV_Materials_to_See_Exploding_Growth.php.

Hansford, Mark, and Jessica Rowson. "Capital Costs Undermine Offshore Wind Farms." New Civil Engineer (NCE), July 2, 2009. http://www.nce.co.uk/news/energy/capital-costs-undermine-offhsore-wind/5204418.article.

Harrington W., R. D. Morgenstern, and P. Nelson, "On the Accuracy of Regulatory Cost Estimates." *Journal of Policy Analysis and Management*, Spring 2000: 297–322.

Harris, Catherine. "Special Report: How Low Will U.S. Auto Sales Go?" *Investment Executive*, April 22, 2009. http://www.investmentexecutive.com/client/en/News/Detail News.asp?Id=49055&IdSection=147&cat=147.

Harvey, Scott M., Bruce M. McConihe, and Susan L. Pope. "Analysis of the Impact of Co-ordinated Electricity Markets on Consumer Electricity Charges." Draft, LECG, November 20, 2006.

Hawkins, David G. Testimony before the Subcommittee on Energy and Environment, "The Future of Coal under Climate Legislation." March 10, 2009.

Haq, Zia. "Biomass for Electricity Generation." Energy Information Administration, May 13, 2002.

Hedman, Bruce. "CHP Market Status." Presented at Gulf Coast Roadmapping Workshop, April 26, 2005.

Hempstead, Jim. "New Nuclear Generating Capacity: Potential Credit Implications for U.S. Investor Owned Utilities." Special Comment, Report Number 109152, *Moody's Corporate Finance*, May 2008.

Hesmondhalgh, Serena, Toby Brown, and David Robinson. "EU Climate and Energy Policy for 2030 and the Implications for Carbon Capture and Storage." *The Brattle Group* for ALSTOM Power Systems, March 2009.

Hequet, Marc. "Damless Hydro Power." *Public Power*, November/December 2008.

Highfill, Darren R., and Vishant Shah. "A Voice for Smart-Grid Security." *Public Utilities Fortnightly*, July 2009: 24–31.

Hledik, Ryan. "How Green Is the Smart Grid?" *Electricity Journal*, April 2009: 29–41.

Holly, Chris. "IPCC Chief Raps Obama's Near-Term Climate Targets." *Energy Daily*, January 21, 2009.

Hogan, William W. "Competitive Electricity Market Design: A Wholesale Primer." December 17, 1998. http://www.hks.harvard.edu/fs/whogan/empr1298.pdf.

———. "Providing Incentives for Efficient Demand Response." October 29, 2009. http://www.hks.harvard.edu/fs/whogan/Hogan_Demand_Response_102909.pdf.

———. "WEPEX: Building the Structure for a Competitive Electricity Market." Presented at the FERC Technical Conference Concerning WEPEX, Washington, DC, August 1, 1996.

Hotinski, Roberta. "Stabilization Wedges: A Concept & Game." Carbon Mitigation Initiative, Princeton Environmental Institute, January 2007.

"Implementing California's Loading Order for Electricity Resources." Staff Report, California Energy Commission, CEC-400-2005-043, July 2005.

"Increasing Efficiency: New PV Cell Conversion Efficiency of 18.9% Represents the World's Highest for a Second Year Running." Mitsubishi Electric CSR. http://global .mitsubishielectric.com/company/csr/ecotopics/pv/cell/index.html.

Insull, Samuel. *Central Station Electric Service.* (Chicago: Privately Printed) 1915.

"Integrating Locationally—Constrained Resources into Transmission Systems: A Survey of U.S. Practices." WIRES (Working group for Investment in Reliable & Economic Electric Systems) and CRA International, October 2008.

"Integrating Wind Power Into the Electric Grid." National Conference of State Legislatures, June 2009.

International Atomic Energy Agency (IAEA). "Workforce Planning for New Nuclear Power Programmes." Draft, Presented at IAEA's Technical Meeting on Workforce Planning to Support New Nuclear Power Programmes, Vienna, Austria, March 31–April 2, 2009.

———. "The Nuclear Power Industry's Ageing Workforce: Transfer of Knowledge to the Next Generation." IAEA-TECDOC-1399, Vienna, Austria, June 2004.

International Energy Agency (IEA). "IEA Energy Efficiency Policy Recommendations to the G8 2007 Summit." Heiligendamm, June 2007.

———. "Energy Efficiency Policy Recommendations." 2008(b).

———. "Energy Technology Perspectives 2008: Scenarios and Strategies to 2050." 2008(a).

Interstate Renewable Energy Council. "Map of State Net Metering Rules." August 2009.

"Introduction to Interoperability and Decision-Maker's Interoperability Checklist, Version 1.0." Gridwise Architecture Council, April 2007.

"IOACTIVE Verifies Critical Flaws in Next Generation Energy Infrastructures." IOACTIVE Press Release, March 23, 2009.

"ISO New England Probes Mid-Jan Rolling Blackout Warnings." Dow Jones Energy Services, March 2, 2004.

Izzo, Ralph. "Climate Change: Taking Hints, Taking Actions." Keynote remarks at Clean Air–Cool Planet: Global Warming and Energy Solutions Conference, October 12, 2007.

———. "Innovation: The Future of Energy." Presented at Silberman College of Business and Rothman Institute of Entrepreneurial Studies, Fairleigh Dickinson University, November 5, 2008.

Jaczko, Gregory B. (NRC). Letter to The Honorable Thomas R. Carper, Chairman, Subcommittee on Clean Air and Nuclear Safety, U.S. Senate regarding report on status of its licensing and other regulatory issues, May 15, 2009.

Jaffe, Adam, Richard G. Newell, and Robert Stavins. "Technological Change and the Environment." National Bureau of Economic Research, October 2000.

———. "Energy-Efficient Technologies and Climate Change Policies: Issues and Evidence." John F. Kennedy School of Government Working Paper, December 1999.

Jamison, Mark A., et al. "Disbanding the Maryland Public Service Commission." Case No. 2006-2, PURC, July 2006.

Johnson, Claire B. "Municipal Energy Financing." U.S. Department of Energy, June 1, 2009.

Johnson, Keith. "Ill Winds Blow for Clean Energy: Cheap, and Abundant, Natural Gas Diminishes Alternative Projects' Appeal." *Wall Street Journal*, July 9, 2009.

"Joint Coordinated System Plan." http://www.JCSPStudy.org.

Jones, John, et al. "Electric Utilities: A New Operating Model." Booz Allen Hamilton, 2002.

Jones, Kevin. "Battle of the Currents." *Electrical Apparatus*, October 1, 2003. *HighBeam Research*. http://www.highbeam.com/doc/1P3-433184831.html. (February 9, 2009).

Jonnes, Jill. *Empires of Light*. (New York: Random House) 2003.

Joskow, Paul L. "California's Electricity Crisis." *Oxford Review of Economic Policy*, 2001: 365–388.

———. "Regulatory Failure, Regulatory Reform and Structural Change in the Electric Power Industry." *Brookings Papers on Microeconomic Activity*; Special Issue, 1989: 125–200.

———. "Vertical Integration." *Handbook of New Institutional Economics*. (Boston: Kluwer) 2003.

———. "Vertical Integration." Chapter XX in *Issues in Competition Law and Policy*, (Section of Antitrust Law), August 6, 2008.

Joskow, Paul L., and Donald B. Marron. "What Does a Negawatt Really Cost? Further Thoughts and Evidence." *Electricity Journal*, July 1993: 14–26.

Joskow, Paul L., and Richard Schmalensee. *Market for Power: An Analysis of Electrical Utility Deregulation*. (Cambridge: MIT Press) 1998.

Kahn, Michael, and Loretta Lynch. "California's Electricity Options and Challenges Report to Governor Gray Davis." California Public Utilities Commission, August 2, 2000.

Kaplan, Stan Mark. "Electric Power Transmission: Background and Policy Issues." Congressional Research Services (CRS) Report for Congress, April 14, 2009.

Kats, Greg, and David Carey. "Upgrading America's Homes Comprehensive Residential Energy Upgrade Financing." May 2009.

Kearney, D. "Assessment of Thermal Energy Storage for Parabolic Trough Solar Power Plants." Presented at NREL's TroughNet—2006 Parabolic Trough Technology Workshop, Incline Village, Nevada, February 14, 2006.

Kelderman, Eric. "States Pull the Plug on Electricity Deregulation." Stateline.org, July 21, 2005. http://www.stateline.org/live/ViewPage.action?siteNodeId=136&languageId=1&contentId=44242.

Kelly, Bruce. "Large Plant Studies." Presented at NREL's TroughNet—2006 Parabolic Trough Technology Workshop, Incline Village, Nevada February 14, 2006.

Kelly, John. "EMRI: Do Competition and Electricity Mix?" American Public Power Association, 2007.

Kelly, Susan, and Diane Moody. "Wholesale Electric Restructuring: Was 2004 the "Tipping Point"?" *Electricity Journal*, March 2005: 11–18.

Kelly, Suedeen. "Federal Legislation and Policy Update." Federal Energy Regulatory Commission. Presented at Pennsylvania Public Utility Law Conference, Harrisburg, PA, May 7, 2009.

Kiesling, L. Lynne. "Project Energy Code—Markets, Technology and Institutions: Increasing Energy Efficiency through Decentralized Coordination." *Econalign*, Series Issue 2, February 2009.

King, Tom. "Transforming the Electric Grid." Oak Ridge National Laboratory, January 6, 2005.

Kirby, Brendan, and Michael Milligan. "Facilitating Wind Development: The Importance of Electric Industry Structure." *Electricity Journal*, April 2008: 40–54.

Klein, Peter G. "The Make-or-Buy Decision: Lessons from Empirical Studies." Working Paper No. 2004-07, CORI, April 2004. http://cori.missouri.edu/wps.

Kleit, Andrew N., and Dek Terrell. "Measuring Potential Efficiency Gains from Deregulation of Electricity Generation: A Bayesian Approach." *Review of Economics and Statistics*, August 2001: 523–530.

Koning, Fabian. "Renewable Energy: Feeling the Heat." *Financial Times*, June 3, 2009.

Krauss, Clifford. "Tightened Codes Bring a New Enforcer, the Energy Inspector." *New York Times*, July 18, 2009.

Kroposki, B., et al. "Renewable Systems Interconnection: Executive Summary." Technical Report NREL/TP-581-42292, NREL, February 2008.

Kwoka, John E. *Power Structure: Ownership, Integration, and Competition in the U.S. Electricity Industry*. (Boston: Kluwer Academic) 1996.

———. "Restructuring the U.S. Electric Power Sector: A Review of Recent Studies." Prepared for the American Public Power Association, November 2006.

———. "Vertical Economies in Electric Power: Evidence on Integration and Its Alternatives." *International Journal of Industrial Organization*, May 2002: 653–671.

Kwoka, John, Sanem Ozturk, and Michael Pollitt. "Divestiture Policy and Operating Efficiency in U.S. Electric Power Distribution." EPRG Working Paper 0819, July 2008.

Krewitt, Wolfram. "Future Technology Options for Electricity Generation: Technical Trends, Costs and Environmental Performance." Presented at Needs Conference on External Costs of Energy Technologies, Brussels, February 16, 2009.

Kühn, Kai-Uwe, and Matilde Machado. "Bilateral Market Power and Vertical Integration in the Spanish Electricity Spot Market." CEPR Discussion Papers, 2004.

Kurt, Kelly. "California Power Crunch Sends Deregulation Shockwaves." *Oklahoma City Journal Record*, 2001. http://findarticles.com/p/articles/mi_qn4182/is_/ai_n10142982.

Kushler, Martin, Dan York, and Patti Witte. "Meeting Aggressive New State Goals for Utility-Sector Energy Efficiency: Examining Key Factors Associated with High Savings." American Council for an Energy-Efficient Economy (ACEEE), U091, March 2009.

Laitner, J. A. "Skip," C. Poland-Knight, V. McKinney, and K. Ehrhardt-Martinez, "Semiconductor Technologies: The Potential to Revolutionize U.S. Energy Productivity." American Council for an Energy-Efficient Economy (ACEEE), Report E094, May 2009.

Lazzaro, Joseph. "Are Speculators Pushing Up Oil Again?" *Daily Finance*, April 23, 2009. http://www.dailyfinance.com/story/are-speculators-pushing-up-oil-again/1525943/.

Lesser, Jonathan, and Nicholas Puga. "PV vs. Solar." *Public Utilities Fortnightly*, July 2008: 17–20, 27.

Letendre, Steven, Paul Henholm, and Peter Lilienthal. "Plug-In Hybrid and All-Electric Vehicles: New Load, or New Resource?" *Public Utilities Fortnightly*, December 2006. http://www.fortnightly.com/display_pdf.cfm?id=12012006_Cars.pdf.

Levine, Mark D., and Diana Ürge-Vorsatz. "IPCC WGIII Assessment Report: Chapter 6. Mitigation Options in Buildings." November 1, 2007.

Lobsenz, George. "UK Moves Forward on Seabed 'Socket' for Wave Energy." *Energy Daily*, July 28, 2009: 1, 3.

Lohr, Steve. "Digital Tools Help Users Save Energy, Study Finds." *New York Times*, January 10, 2008.

Lovins, Amory, et al. "Energy Strategy: The Road Not Taken?" *Foreign Affairs*. October, 1976: 5–15.

———. *Soft Energy Paths: Toward a Durable Peace*. (New York: HarperCollins) 1979.

———. *Small Is Profitable*. (Snowmass: Rocky Mountain Institute) 2002.

Lynd, Robert S., and Helen M. Lynd. *Middletown*. (New York: Harcourt, Brace) 1929.

MaCabrey, Jean-Marie. "Soaring Electricity Use by New Electronic Devices Imperils Climate Change Efforts." *Climate Wire*, May 14, 2009.

Mall, Amy, Sharon Buccino, and Jeremy Nichols. "Drilling Down: Protecting Western Communities from the Health and Environmental Effects of Oil and Gas Production." Natural Resources Defense Council (NRDC), October 2007.

Mansur, Erin T. "Measuring Welfare in Restructured Electricity Markets." *Review of Economics and Statistics*, May 2008: 369–386.

———. "Upstream Competition and Vertical Integration in Electricity Markets." *Journal of Law and Economics*, February 2007: 125–156.

Markiewicz, Kira, Nancy Rose, and Catherine Wolfram. "Has Restructuring Improved Operating Efficiency at US Electricity Generating Plants?" CSEM WP 135, UCEI, July 2004.

Marshall, Christa. "Policy: Carbon Capture and Storage Moves to Center Stage in Cap-and-Trade Debate." *Climate Wire*, June 9, 2009. http://www.eenews.net/climate wire/2009/06/09/2/.

McClure, George. "Energy Fixes: Smart Grid, Nuclear Plants." Today's Engineer, November 2008. http://www.todaysengineer.org/2008/Nov/energy.asp.

McCullough, Robert. "California Electricity Price Spikes: Factual Evidence." January 15, 2003. http://www.mresearch.com/pdfs/76.pdf.

McCullough, Robert. Memorandum to McCullough Research Clients, "C66 and the Artificial Congestion of California Transmission in January 2001." November 29, 2002.

McKinsey & Company. "Reducing U.S. Greenhouse Gas Emissions: How Much at What Cost?" December 2007.

———. "Pathways to a Low-Carbon Economy: Version 2 of the Global Greenhouse Gas Abatement Cost Curve." 2009: 15.

"Meeting Projected Coal Production Demands in the USA: Upstream Issues, Challenges and Strategies." The National Commission on Energy Policy (NCEP), December 2008.

"Methodology and Forecast of Long Term Avoided Costs for the Evaluation of California Energy Efficiency Programs." Prepared for CPUC, October 25, 2004.

Metz, Bert, et al. (eds.). *Carbon Dioxide Capture and Storage*. Report by the Intergovernmental Panel on Climate Change, (Cambridge: Cambridge University Press) 2005.

Meyers, Stephen, James McMahon, and Michael McNeil. "Realized and Prospective Impacts of U.S. Energy Standards for Residential Appliances: 2004 Update." LBNL-56417, Lawrence Berkeley National Laboratory, May 2005.

Michaels, Robert J. "Vertical Integration and the Restructuring of the U.S. Electricity Industry." *Policy Analysis*, (572), July 13, 2006.

Miller, Christopher G. "The Future of the Grid: Proposal for Reforming National Transmission Policy." Testimony before the Subcommittee on Energy and the Environment, Committee on Energy and Commerce, U.S. House of Representatives, June 12, 2009.

Moody, Diane. "The Use—and Misuse—of Statistics in Evaluating the Benefits of Restructured Electricity Markets." *Electricity Journal*, March 2007: 57–62.

Morgan & Company, Inc. "North American Vehicle Forecast/Global Outlook." Automotive Forecast Services, 2008. http://www.morgancom.com/vehicleforecasts.htm.

Morrison, Jay A. "The Clash of Industry Visions." *Electricity Journal*, January/February 2005: 14–30.

Moss, Diana L., and Peter Fox-Penner. "Introduction." *Network Access, Regulation and Antitrust.* (New York: Routledge) 2005: 1–9.

Motion of the California ISO and EOB, Docket No. El00-95-000 et al., March 3, 2003: 5.

Mudd, Michael J. Testimony before the U.S. Committee on Commerce, Science and Transportation, April 9, 2008.

Nadel, Steven. "Success with Energy Efficiency Resource Standards." ACEEE, January 2009(b).

———. "Energy Policy Trends: Looking Both Inside and Beyond 'the Beltway.'" American Council for an Energy-Efficiency Economy, January 2009(a).

———. Testimony before the Senate Energy Committee, Hearing on Energy Efficiency Resource Standards, April 22, 2009(c).

Nadel, Steven, and Miriam Pye. "Appliance and Equipment Efficiency Standards: Impacts by State." ACEEE report, 1996. http://www.aceee.org/pubs/a964.htm.

"NAFO praises biomass/carbon provisions in energy act." *Macon County News*, July 9, 2009. http://www.maconnews.com/index.php?option=com_content&task=view&id=5024&Itemid=116.

Nakashima, Ellen, and Steve Mufson. "Hackers Have Attacked Foreign Utilities, CIA Analyst Says." *Washington Post*, January 19, 2008.

National Academy of Engineering. "Greatest Engineering Achievements of the 20th Century." 2009. http://www.greatachievements.org/.

"National Action Plan for Energy Efficiency Vision for 2025: A Framework for Change." November 2008. http://www.epa.gov/cleanenergy/energy-programs/napee/resources/vision2025.html.

National Association of Regulatory Utility Commissioners (NARUC). "Decoupling for Electric & Gas Utilities: Frequently Asked Questions (FAQ)." September 2007.

National Commission on Energy Policy. "Forging the Climate Consensus: Managing Economic Risk in a Greenhouse Gas Cap-and-Trade Program." July 2009.

National Grid. "General Information Pack." June 2009: 61. http://www.nationalgrid.com/corporate/Investor+Relations/Presentations+and+webcasts/2009-10/General+Information+Pack+17Jun09.htm.

National Renewable Energy Laboratory (NREL). "Renewable Energy Cost Trends." 2005. http://www.nrel.gov/analysis/docs/cost_curves_2005.ppt.

———. "Renewable Systems Interconnection." Executive Summary, February 2008.

Neenan, Bernard, and Jiyong Eom, "Price Elasticity of Demand for Electricity: A Primer and Synthesis." Electric Power Research Institute, 2008.

Nemoto, Jiro, and Mika Goto. "Technological Externalities and Economies of Vertical Integration in the Electric Utility Industry." *International Journal of Industrial Organization*, January 2004: 67–81.

Neuhoff, Karsten. "Large Scale Deployment of Renewables for Electricity Generation." CMI Working Paper 59, March 14, 2006.

Neville, Angela, J. D. "Prevailing Winds: Trends in U.S. Wind Energy." *Power*, December 1, 2008. http://www.powermag.com/renewables/wind/Prevailing-winds-Trends-in-U-S-wind-energy_1573.html.

"New PV Cell Conversion Efficiency of 18.9% Represents the World's Highest for the Second Year Running." Mitsubishi Electric, 2009. http://global.mitsubishielectric.com/company/csr/ecotopics/pv/cell/index.html.

"New York Shopping Sees Growth in Every Metric." *Restructuring Today*, July 1, 2008.

Newell, Samuel, and Ahmad Faruqui. "Dynamic Pricing: Potential Wholesale Market Benefits in New York State." *The Brattle Group*, August 2009.

Newman, Sam, et al. "Accelerating Solar Power Adoption: Compounding Cost Savings across the Value Chain." Rocky Mountain Institute, 2009. http://www.rmi.org/rmi/Library/2009-03_AcceleratingSolarPowerAdoption.

Niles, Raymond C. "Property Rights and the Crisis of the Electric Grid." *Objective Standard*, Vol. 3, Summer 2008.

North American Electric Reliability Corporation (NERC). "2008 Long-Term Reliability Assessment 2008–2017." October 2008.

————. "Accommodating High Levels of Variable Generation." April 2009.

————. "Estimated 2017 Summer Margins, Resources and Demands." 2008 Long Term Reliability Assessment, October 2008: Tables 13a and 13e.

————. "Special Report: Electric Industry Concerns on the Reliability Impacts of Climate Change Initiatives." November 2008: Chap. 8.

North, Gary. "Enron, Spawn of Business Journalism." Column on LewRockwell.com, February 6, 2002. http://www.lewrockwell.com/north/north92.html.

Nourai, Ali. "Store and Deliver." *Power Energy Magazine*, Q.4, 2009. http://www.nextgenpe.com/article/Issue-7/Renewables/Store-and-Deliver.

"Nuclear Energy for the Future: Required Research and Development Capabilities." Battelle, September 2008.

Nuclear Energy Institute (NEI). "Policies That Support New Nuclear Power Plant Development." Fact Sheet, January 2009. http://www.nei.org/resourcesandstats/documentlibrary/newplants/factsheet/policiessupportnewplantdevelopment.

"Nuclear Energy Outlook 2008—Executive Summary." Organisation for Economic Co-operation and Development (OECD). Moulineaux, France, November 28, 2008.

"Nuclear Plant Construction Poses Risks to Credit Metrics, Ratings." Global Credit Research Announcement, *Moody's Investors Service*, June 2, 2008.

"Nuclear Power in the USA." The World Nuclear Association, July 2009. http://www.world-nuclear.org/info/inf41.html.

Nye, David E. *Electrifying America: Social Meanings of a New Technology*. Cambridge: MIT Press, 1995.

O'Malley, Martin. "Taking Control of Our Energy Future." Keynote address given to the

Maryland Association of Counties (MACo) Annual Summer Conference, Ocean City, MD, August 16, 2008.

Oatman, Gene, and Dave Crudele. "The Galvin Path to Perfect Power—A Technical Assessment." The Galvin Electricity Initiative, January 2007.

Office of Trade and Economic Development. Energy Division. "Issues and Analyses for the Washington State Legislature." *2000 Biennial Energy Report*, January 2001.

Osborn, D., and Z. Zhou. "Transmission Plan Based on Economic Studies." Midwest ISO, April 2008.

"PG&E and BrightSource Sign Record Solar Power Deal." BrightSource Press Release, May 13, 2009.

Pacala, S., and R. Socolow. "Stabilization Wedges: Solving the Climate Problem for the Next 50 Years with Current Technologies." *Science*, August 13, 2004: 968–972.

Pachauri, R. K., and A. Reisinger (eds.). "Climate Change 2007: Synthesis Report." International Panel on Climate Change, November 2007.

Pagnamenta, Robin. "France Imports UK Electricity as Plants Shut." *TimesOnLine*, July 3, 2009. http://business.timesonline.co.uk/tol/business/industry_sectors/utilities/article6626811.ece.

Paidipati, J., et al. "Rooftop Photovoltaics Market Penetration Scenarios." NREL/SR-581-42306, NREL, February 2008.

Parmesano, Hethie. "Rate Design Is the No. 1 Energy Efficiency Tool." *Electricity Journal*, July 2007: 18–25.

Paul, Anthony. "Electricity Consumption Efficiency a Top-Down Assessment of Potential." Electricity and Environment Program, Resources for the Future. Presented at the Behavior, Energy, and Climate Change Conference (BECC) in Sacramento CA, November 17, 2008. http://piee.stanford.edu/cgi-bin/docs/behavior/becc/2008/presentations/17-1E-02-Electricity_Consumption_Efficiency_-_A_Top-Down_Assessment_of_Potential.pdf.

"Performance-Based Rates for U.S. Electric Utilities: A 2007 Status Report." Newton-Evans Research Company, Inc., June 2007.

Pernick, Ron, and Clint Wilder. "Utility Solar Assessment (USA) Study: Reaching Ten Percent Solar by 2025." Clean Edge, June 2008.

Peters, Mark. "Retail Electricity Suppliers Battle Utilities for Customers." *Wall Street Journal Online*, August 31, 2009.

Petrill, Ellen. "Creating a Secure Low-Carbon Future . . . A Framework for Action." ASERTTI, October 16, 2008.

"Pew Center Summary of H.R. 2454: American Clean Energy and Security Act of 2009 (Waxman–Markey)." Pew Center, 2009. http://www.pewclimate.org/acesa.

Pfeifenberger, J. P., G. N. Basheda, and A. C. Schumacher. "Restructuring Revisited." *Public Utilities Fortnightly*, June 2007: 68. http://www.fortnightly.com/pubs/06012007_RestructuringRevisited%20.pdf.

Phillips, Charles F., Jr. *The Regulation of Public Utilities.* (Arlington, VA: Public Utilities Report) 1993: 428–429.

Piedmont Environmental Council, v. Federal Energy Regulatory Commission, No. 07-1651 et al., U.S. Court of Appeals 4th Cir. 2008, 2009. http://pacer.ca4.uscourts.gov/opinion.pdf/071651.P.pdf.

Pinderhughes, Raquel, Ph.D. "Green Collar Jobs: An Analysis of the Capacity of Green Businesses to Provide High Quality Jobs for Men and Women with Barriers to Employment." for the City of Berkeley Office of Energy and Sustainable Development, December 5, 2007.

Platt, Harold L. *The Electric City.* (Chicago: University of Chicago Press), 1991: 76.

Pollin, Robert, et al., "Green Recovery: A Program to Create Good Jobs and Start Building a Low-Carbon Economy." Center for American Progress, September 2008.

"Point Carbon EUA OTC Assessment." Historic Prices, Point Carbon, 2009. http://www.pointcarbon.com/new/historicprices.

Pope, Susan L. "California Electricity Price Spikes: An Update on the Facts." Harvard Electricity Policy Group, December 9, 2002.

"Potential Gas Committee Reports Unprecedented Increase in Magnitude of U.S. Natural Gas Resource Base." Press Release, Colorado School of Mines, June 18, 2009.

"Potential Supply of Natural Gas in the United States." Potential Gas Committee, Colorado School of Mines, December 31, 2009.

Price, H. "Cooling for Parabolic Trough Power Plants." NREL, presented at 2006 Parabolic Trough Technology Workshop, Incline Village, NV, February 14, 2006.

"Quantifying Demand Response Benefits in PJM." Prepared by *The Brattle Group*, January 29, 2007.

Rawson, Mark, and John Sugar. "Distributed Generation and Cogeneration Policy Roadmap for California." Staff Report, CEC-500-2007-021, California Energy Commission, March 2007: 1.

"Recommendations for Improving Enforcement and Protecting Consumers in Deregulated Energy Markets." Attorney General's Energy White Paper, State of California, April 2004: 2.

"Readiness of the U.S. Nuclear Workforces for 21st Century Challenges." A report of the APS Panel on Public Affairs, American Physics Society. June 2008.

"Redefining Our Boundaries, 2008 Summary Annual Report." Duke Energy, 2009

Reed, David. "Fouling Our Own Net." *IEEE Spectrum*, February 2009: 18–19.

"Regulatory Issues for Direct-Use Geothermal Resource Development in Oregon." Kevin Rafferty, ed., Oregon Water Resources Department. http://geoheat.oit.edu/pdf/tp114.pdf.

Reitzes, James D., et al. "Review of PJM's Market Power Mitigation Practices in Comparison to Other Organized Electricity Markets." *The Brattle Group*, September 14, 2007.

"Report to Congress on Competition in the Wholesale and Retail Markets for Electric

Energy." Draft, The Electric Energy Market Competition Task Force and FERC, June 5, 2006.

"Retrofitting of Coal-Fired Power Plants for CO_2 Emissions Reductions." An MIT Energy Initiative Symposium, Massachusetts Institute of Technology, March 23, 2009.

Rocky Mountain Institute. "Micropower Database: How Distributed Renewables and Cogeneration are Beating Nuclear Power Stations." May 3, 2008.

Romm, Joseph. "Is 450 ppm (or less) Politically Possible? Part 2: The Solution." Center for American Progress, April 22, 2008. http://climateprogress.org/2008/04/22/is-450-ppm-or-less-politically-possible-part-2-the-solution/.

———. "The Self-Limiting Future of Nuclear Power." Center for American Progress Action Fund, June 2008.

———. "The Staggering Cost of Nuclear Power: Part One in a Series on a New Nuclear Cost Study." Center for American Progress, January 5, 2009(a). http://american progress.org/issues/2009/01/nuclear_power.html/.

———. The Hype about Hydrogen. (Washington, DC: Island Press) 2004, 2005.

———. "Warning to Taxpayers, Investors: Nukes May Become Troubled Assets, Part Two in a Series on a New Nuclear Cost Study." Center for American Progress, January 7, 2009(b). http://americanprogress.org/issues/2009/01/nuclear_power_part2.html/.

Rosen, Richard. "Regulating Power: An Idea Whose Time Is Back." *American Prospect*, 2002: 22.

Rowland, Kate. "Get Smart." *EnergyBiz Insider*, June 29, 2009. http://www.energycentral.com/articles/energybizinsider/ebi_detail.cfm?id=708.

Russo Marsh & Rogers, Inc. (RMR) "The Corruption of Power." http://www.geoff metcalf.com/loretta_20010329.html.

Sant, Roger W. "Coming Markets for Energy Services." *Harvard Business Review*, May–June 1980: 6–20.

Sappington, David, and others. "The State of Performance-Based Regulation in the U.S. Electric Utility Industry." *Electricity Journal*, October 2001: 71–79.

Sassoon, David. "House Testimony Undermines Wisdom of Massive Electric Grid Expansion." *SolveClimate*, June 17, 2009. http://solveclimate.com/blog/2009061/house-testimony-undermines-wisdom-massive-electric-grid-expansion.

———. "Transmission Superhighway On Track to Carry Cheap, Dirty Coal Power to Northeast." *SolveClimate*, March 24, 2009. http://solveclimate.com/blog/20090324/transmission-superhighway-track-carry-cheap-dirty-coal-power-northeast.

Satchwell, Andy. "Smart Grid Barriers and a Collaborative Solution." June 24, 2008. http://www.aceee.org/conf/08ss/08sstoc.pdf.

Sathaye, Jayant, and Amol Phadke. "Representing Energy Demand in Large Scale Models: Baselines, Market Failures, and Non-Energy Benefits." Presented at Energy Modeling Forum, Washington, DC, February 5, 2008.

Sauer, Amy, and Fredric Beck. "Loan Guarantee Provisions in the 2007 Energy Bills:

Does Nuclear Power Pose Significant Taxpayer Risk and Liability?" Issue Brief, Environmental and Energy Study Institute, October 30, 2007.

Savage, John. "Western Renewable Energy Zone (WREZ) Project." Oregon Public Utility Commission, April 8, 2009.

"Scenario Planning." Wikipedia.org, accessed July 23, 2009.

Schlegel, Jeff. "Energy Efficiency and Global Warming: What Must Be Done." Presentation at the ACEEE Conference: Energy Efficiency as a Resource, October 2, 2007.

Schlosser, Julie. "How Duke Energy's CEO Got Started." *Fortune Magazine*, August 11, 2009. http://money.cnn.com/2009/08/10/magazines/fortune/duke_energy_jim_rogers.fortune/index.htm.

Schwartz, Ariel. "First Commercial Hydrokinetic Turbine Installed In US." CleanTechnica, December 26, 2008. http://cleantechnica.com/2008/12/26/first-commercial-hydrokinetic-turbine-installed-in-us/.

Schweppe, Fred, et al. *Spot Pricing of Electricity*. (Boston: Kluwer) 1988.

"Sector Profiles of Significant Large CHP Markets." Energy and Environmental Analysis, Inc., March 9, 2004.

Seltzer, Larry. "Drive By Blackouting." PCMag.com, Blog, June 15, 2009.

"Sequim, Washington." City-Data.com. http://www.city-data.com/housing/houses-Sequim-Washington.html.

Severance, Craig A. "Business Risks to Utilities as New Nuclear Power Costs Escalate." *Electricity Journal*, May 2009: 112–120.

———. "Business Risks and Costs of New Nuclear Power." *Climate Progress*, June 5, 2009. http://climateprogress.org/wp-content/uploads/2009/01/nuclear-costs-2009.pdf.

Sheffrin, Anjali. Anjali Sheffrin to Market Issues/ADR Committee of the California ISO, memorandum, Market Analysis Report, June 16, 2001(a).

———. Anjali Sheffrin to CAISO Board of Governors, memorandum, Market Analysis Report for September 2001, October 9, 2001(b).

———. Anjali Sheffrin to Market Issues/ADR Committee of the California ISO, memorandum, January 13, 2000, Market Analysis Report.

Sherwood, Larry. "U.S. Market Trends: Solar and Distributed Wind." Interstate Renewable Energy Council, September 24, 2007.

———. "U.S. Solar Market Trends 2008." Interstate Renewable Energy Council, 2009.

Shipley, Anna, et al. "Combined Heat and Power: Effective Energy Solutions for a Sustainable Future." ORNL/TM-2008/224, Oak Ridge National Laboratory, December 1, 2008.

Shy, Oz. *The Economics of Network Industries*. (Cambridge: Cambridge University Press) 2001.

Siddiqui, Omar, K. Parmenter, and P. Hurtado. "The Green Grid: Energy Savings and Carbon Emissions Reduced Enabled by a Smart Grid." EPRI Technical Update 1016905, June 2008.

Silberglitt, Richard, Emile Ettedgui, and Anders Hove. *Strengthening the Grid: Effects of High-Temperature Superconducting Power Technologies on Reliability, Power Transfer Capacity, and Energy Use.* (RAND) 2002.

Sims, Ralph E. "Hydropower, Geothermal, and Ocean Energy." *MRS Bulletin,* April 2008: 389–395. http://www/mrs.org/s_mrs/bin.aps?CID=12527&DID=206495.

Sioshansi, Fereidoon P. "Carbon Constrained: The Future of Electricity Generation." *Electricity Journal,* June 2009: 64–74.

Slack, Kara. "U.S. Geothermal Power Production and Development Update." Geothermal Energy Association, August 2008.

"Smart Grid Standards Adoption: Utility Industry Perspective." Prepared for Smart Grid Utility Executive Working Group and Open SG Subcommittee, May 27, 2009.

"Smart Meters—Got Bugs?" Smart Grid Stimulus, June 17, 2009. http://www.smartgrid stimulus.com/artman/publish/Watch_List_News-Digest/FOAs_for_Big_Smart_ Grid_Projects_a_No-Show_So_Far-687.html.

Smead, Richard G., and Gordon B. Pickering. "North American Natural Gas Supply Assessment." Navigant Consulting, July 4, 2008.

Smith, Rebecca. "Smart Meter, Dumb Idea?" *Wall Street Journal,* April 27, 2009.

———. "Less Demand, Same Great Revenue." *Wall Street Journal,* February 8, 2009

———. "U.S. Foresees a Thinner Cushion of Coal." *Wall Street Journal,* June 8, 2009.

Solar Energy Industries Association (SEIA). "U.S. Solar Industry Year in Review 2008." Executive Summary, April 2009. http://www.seia.org/galleries/pdf/2008_Year_in_ Review-small.pdf.

Sovacool, Benjamin K., and Charmaine Watts. "Going Completely Renewable: Is It Possible (Let Alone Desirable)?" *Electricity Journal,* May 2009, Vol. 22, (4): 95–111.

Steinhurst, William, et al. "Energy Portfolio Management: Tools and Resources for State Public Commissions." Prepared by Synapse Energy Economics, for NARUC, October 2006.

Strader, Jim, and Paula C. Squires. "Electric Deregulation." *Virginia Business,* August 2004.

"Strategies to Meet Possible 33% RPS in California Hatched." *Restructuring Today,* July 22, 2009.

Stoft, Steven. "Power System Economics: Designing Markets for Electricity." (New York: Wiley-IEEE Press) 2002.

Sudarshan, Anant, and James Sweeney. "Deconstructing the 'Rosenfeld Curve.'" Stanford University. July 1, 2008. http://piee.stanford.edu/cgi-in/docs/publications/ Deconstructing_the_Rosenfeld_Curve.pdf.

Sullivan, Patrick, et al. "Comparative Analysis of Three Proposed Federal Renewable Electricity Standards." NREL/TPL-6A2-45877, National Renewable Energy Laboratory, May 2009.

Summers, Adam B. "Call for Real Electricity Deregulation." *Ventura County Star,* October 25, 2005. http://reason.org/news/show/122806.html.

Sutherland, Ronald J. "Market Barriers to Energy-Efficiency Investments." *Energy Journal*, July 1, 1991: 15–34.

Sweeney, James L. *The California Electricity Crisis.* (Hoover Institution Press) July 2002.

"TVA Goes to War." *TVA Heritage.* The Tennessee Valley Authority, Oak Ridge, TN., 2009. http://www.tva.gov/heritage/war/index.htm.

Taber, John, Duane Chapman, and Tim Mount. "Examining the Effects of Deregulation on Retail Electricity Prices." WP2005-14, February 2006.

"Tackling Climate Change in the U.S.: Potential Carbon Emissions Reductions from Energy Efficiency and Renewable Energy by 2030." Edited by Charles F. Kutshcer, American Solar Energy Society, January 2007.

Talukdar, Sarosh N., et al. "Cascading Failures: Survival versus Prevention." *Electricity Journal*, November 2003: 25–31.

Taylor, Jerry, and Peter Van Doren. "Myth Five—Price Signals Are Insufficient to Induce Efficient Energy Investments." *Energy and American Society: Thirteen Myths.* (New York: Springer) 2007: Chapter 6.

Thaler, Richard H., and Cass R. Sunstein. *Nudge.* (New Haven: Yale University Press) 2008: 6–7.

"The Best CEOs in America." *Corporate Leader Magazine*, November 2007.

"The Competitive Retail Electricity Markets and Customer Protections: Are customers really better off?" Regulatory Compliance Services, March 2009.

"The Elusive Negawatt—Energy Efficiency." *Economist*, May 10, 2008.

"The Energy Imperative: Technologies and the Role of Emerging Companies." Report of the President's Council of Advisors on Science and Technology, November 2006.

"The Energy Imperative: Report Update." Report of the President's Council of Advisors on Science and Technology, November 2008(a).

"The Future of Coal." MIT Study, 2007. http://web.mit.edu/coal/.

The New Palgrave: A Dictionary of Economics. Eatwell, John, Murray Milgate, and Paul Newman, eds. (London: Macmillan, 1987).

Thompson, Ben. "Meet the Maverick." *Power and Energy Magazine*, December 2007: 38–41.

Thompson, Clive. "A Green Coal Baron?" *New York Times*, June 22, 2008.

Thornberg, Christopher F. "Of Megawatts and Men: Understanding the Causes of the California Power Crisis." The UCLA Anderson Forecast, 2002.

Tiernan, Tom. "Cost Allocation Clarity Needed, Parties Say." *Megawatt Daily*, January 14, 2009.

Tierney, Susan F. "Decoding Developments in Today's Electric Industry—Ten Points in the Prism." Analysis Group, October 2007.

Toole, W. R., and Susan Winsor. "Nuclear Workforce Dwindles While Needs for Future Grow." *Aiken Standard*, January 28, 2009. http://www.aikenstandard.com/OpEd Columns/011909-winsor-column.

Tomain, Joseph P. "Building the iUtility." *Public Utilities Fortnightly*, August 2008: 28–33.

Trebing, Harry M. "The Networks as Infrastructure—The Reestablishment of Market Power." *Journal of Economics Issues*, 28, 1994: 379–389.

Tutak, Jocelyn, and James R. Stritholt, "Assessing the Impact of Ecological Considerations on Forest Biomass Projection." Conservation Biology Institute, September 2009. http://www.consbio.org/what-we-do/assessing-the-impact-of-ecological-considerations-on-forest-biomass-projections-for-the-southeastern-u.s.

UCEI (University of California Energy Institute). "Zonal Price and Quantity Data." http://www.ucei.berkeley.edu/.

"U.K. Says New Plants Must Capture Carbon." *Greenwire*, April 23, 2009.

U.S. Census Bureau. "Table 1. Projections of the Population and Components of Change for the United States: 2010 to 2050." Population Division, August 14, 2008.

———. "Summary of Energy Efficiency Provisions and Funding Opportunities—American Recovery and Reinvestment Act of 2009." February 17, 2009.

U.S. Congress. House. "American Clean Energy and Security Act." H.R. 2454, Sec. 111. National Strategy, 111th Congress, 1st sess., June 26, 2009: 86–104.

U.S. Department of Defense (DOD). News Briefing—Secretary Rumsfeld and Gen. Myers, Office of the Assistant Secretary of Defense (Public Affairs), February 12, 2002.

U.S. Department of Energy (DOE). "Secretary Chu Announces Two New Projects to Reduce Emissions from Coal Plants." July 1, 2009. http://www.energy.gov/7559.htm.

———. Smart Grid E-Forums. "Smart Grid: The Value Proposition for Consumers." October 23, 2008(a).

———. Electricity Advisory Committee. "Keeping the Lights on in a New World." Washington, DC, January 2009. http://www.oe.energy.gov/eac.htm.

———. Energy Efficiency and Renewable Energy. "20% Wind Energy by 2030: Increasing Wind Energy's Contribution to U.S. Electric Supply." DOE/GO-102008-2567, Washington, DC, July 2008(b). http://www1.eere.energy.gov/windandhydro/pdfs/41869.pdf.

———. "Cost Trends." Energy Efficiency and Renewable Energy (EERE). http://www.windpoweringamerica.gov/ne_economics_cost.asp.

———. "Growing America's Energy Future: Biomass Program." Energy Efficiency and Renewable Energy (EERE). http://www1.eere.energy.gov/biomass/pdfs/sustainability.pdf.

———. "Renewable Energy Data Book." Energy Efficiency and Renewable Energy (EERE). September 2008(c). http://www1.eere.energy.gov/maps_data/renewable_energy.html.

———. Office of Electricity Delivery and Energy Reliability. "The Smart Grid: An Introduction." 2009. http://www.oe.energy.gov/1165.htm.

———. Office of Electricity Delivery and Energy Reliability. "Summary of Regulations Implementing Federal Power Act Section 216(h)." September 2008. http://www.oe.energy.gov/DocumentsandMedia/Summary_of_216_h__rules_clean(1).pdf.

———. Office of Fossil Energy, National Energy Technology Laboratory. *Carbon Sequestration Atlas of the United States and Canada*, 2nd ed., 2008.

U.S. Department of the Interior (DOI). "Framework for Geological Carbon Sequestration on Public Land." Report to Congress, Washington, DC, June 2009(a).

———. "Secretary Salazar, Senator Reid Announce 'Fast-Track' Initiatives for Solar Energy Development on Western Lands." Press Release, Washington, DC, June 29, 2009(b).

U.S. Environmental Protection Agency. "EPA Analysis of the Waxman–Markey Discussion Draft: The American Clean Energy and Security Act of 2009." Washington, DC, April 20, 2009(a): 4, 30.

———. "EPA Analysis of the American Clean Energy and Security Act of 2009." H.R. 2454 EPA Data Annex—ADAGE & IGEM v2.3, 111th Congress, Washington, DC, June 23, 2009(b). http://www.epa.gov/climatechange/economics/pdfs/HR2454_Analysis.pdf.

———. "Federal Requirements under the Underground Injection Control (UIC) Program for Carbon Dioxide (CO_2) Geologic Sequestration (GS) Wells." Proposed Rule, 40 CFR Parts 144 and 144, *Federal Register*, July 25, 2008.

———. "National Air Pollutant Emission Trends, 1900–1998. Trend Charts." *Summary of National Emission Trends*. (Washington, DC: Environmental Protection Agency) March 2000.

U.S. Geological Survey. "Assessment of Moderate- and High-Temperature Geothermal Resources of the United States." Fact Sheet, 2008.

U.S. Government Accountability Office. "The Department of Energy: Key Steps Needed to Help Ensure the Success of the New Loan Guarantee Program for Innovative Technologies by Better Managing Its Financial Risk." GAO-97-339R, GAO Report, Washington, DC, February 28, 2007.

U.S. House Committee on Commerce. Subcommittee on Energy & Power. "Electricity Regulation—A Vision for the Future: Testimony of Kenneth L. Lay." 104th Congress, May 15, 1996. Available from: Department of Energy. http://management.energy.gov/documents/enron1996.pdf.

U.S. Nuclear Regulatory Commission. "Expected New Nuclear Power Plant Applications." Updated September 28, 2009. http://www.nrc.gov/reactors/new-reactors/new-licensing-files/expected-new-rx-applications.pdf.

U.S. Senate Committee on Energy and Natural Resources. "Bingaman–Specter 'Low Carbon Economy Act' of 2007." http://energy.senate.gov/public/_files/LowCarbon EconomyActTwoPager0.pdf.

United States Clean Heat & Power Association. "Clean Heat and Power Provisions, of the American Clean Energy and Security Act of 2009." http://www.uschpa.org/files/public/WMCHP.pdf.

Vajjhala, Shalini P. "Siting Renewable Energy Facilities: A Spatial Analysis of Promises and Pitfalls." Resources for the Future Discussion Paper, RFF DP 06-34, July 2006.

Vajjhala, Shalini, et al. "Green Corridors: Linking Interregional Transmission Expansion and Renewable Energy Policies." Resources for the Future Discussion Paper, RFF DP 08-06, March 2008: 13.

Ventyx Velocity Suite, 2009.

Vock, Daniel C. "Smart Grid's Growth Now Depends on States." *McClatchy-Tribune Regional News*, March 17, 2009.

Vojdani, Ali. "Smart Integration: The Key to Unlock Demand Response Power from Smart Metering Investments." *Metering International*, (1), 2009: 108–109.

———. "Smart Integration: The Smart Grid Needs Infrastructure That Is Dynamic and Flexible." *IEEE Power & Energy Magazine*, November/December 2008: 71–79.

Wack, Pierre. "Shooting the Rapids." *Harvard Business Review*, November 1985. http://hbr.org/1985/11/scenarios/ar/1.

Wallsten, Scott J. "The R&D Boondoogle." *Regulation*, January 17, 2001: 12–16.

Wasik, John F. *The Merchant of Power*. (New York: Palgrave MacMillan) 2006: 24.

Weare, Christopher. "The California Electricity Crisis: Causes and Policy Options." Public Policy Institute of California, 2003.

Weaver, Jaqueline Lang. "Can Energy Markets Be Trusted? The Effect of the Rise and Fall of Enron on Energy Markets." *Houston Business and Tax Law Journal*, Vol. 4, 2004. http://www.hbtlj.org/v04/v04Weaverar.pdf.

Weiss, Charles, and William B. Bonvillian. Structuring an Energy Technology Revolution. (Cambridge, MA: MIT Press) 2009: 16.

Welch, Mark, and Kieran McLoughlin. "Information Is Power: The Intelligent Utility Network." IBM. http://www-03.ibm.com/industries/utilities/us/detail/resource/Y448128V62075A69.html.

Wellinghoff, Jon, and David Morenoff, "Recognizing the Importance of Demand Response: The Second Half of the Wholesale Electric Market Equation." *Energy Law Journal*, Vol. 28, 2007: 389–419.

Westervelt, Robert. "Experts Alarmed over U.S. Electrical Grid Penetration." SearchSecurity.com, April 8, 2009. http://searchsecurity.techtarget.com/news/article/0,289142,sid14_gci1353208,00.html.

Weston, Frederick. "Customer-Sited Resources and Utility Profits: Aligning Incentives with Public Policy Goals." Presented at U.S. EPA Webinar, August 28, 2008.

Weyant, John P. "EMF Briefing on Climate Policy Scenarios: U.S. Domestic and International Policy Architectures." Overview, updated, June 12, 2009. http://emf.stanford.edu/events/emfbriefing_on_climate_policy_scenarios_us_domestic_and_international_policy_architectures.

"Whirlpool Commits to Smart Grid–Compatible Appliances." *Energy Daily*, May 18, 2009.

Williamson, Oliver E. *Transaction Cost Economics*. (Northhampton, MA: Edward Elgar Publishing) 1975.

———. "Transaction-Cost Economics: The Governance of Contractual Relations." *Journal of Law and Economics*, October 1979: 233–261.

"Wind Power Project Site." Global Energy Concepts, October 2005.

"Wind Power: Wind Technology Today." Community Wind Power Fact Sheet #1, University of Massachusetts at Amherst. http://www.ceere.org/rerl/about_wind/RERL_Fact_Sheet_1_Wind_Technology.pdf.

Winston, Clifford. "Days of Reckoning for Microeconomists." *Journal of Economic Literature*, September 1993: 1263–1289.

Wiser, Ray, Galen Barbose, and Carla Peterman. "Tracking the Sun: The Installed Cost of Photovoltaics in the U.S. from 1998–2007." Lawrence Berkeley National Laboratory, February 2009.

Wiser, Ray, and Mark Bolinger. "2008 Wind Technologies Market Report." Lawrence Berkeley National Laboratory, July 2009.

———. "Annual Report on U.S. Wind Power Installation, Cost, and Performance Trends: 2007." Report Summary, Lawrence Berkeley National Laboratory, May 2008.

Wolak, Frank A. "Report on Redesign of California Real-Time Energy and Ancillary Services Market." October 18, 1999.

———. "What Went Wrong with California's Re-structured Electricity Market? (And How to Fix It)." Presented at the Stanford Institute for Economic Policy Research, February 15, 2001.

Wolak, Frank A., Robert Nordhaus, and Carl Shapiro. "The Competitiveness of the California Energy and Ancillary Service Market." March 9, 2000. http://www.caiso.com/docs/2000/08/01/2000080111213227437.pdf.

Wood, Pat, and Rob Church. "Building the 21st Century Transmission Super Grid: Technical and Political Challenges for Large Scale Renewable Electricity Production in the U.S." American Council on Renewable Energy (ACORE), April 2009.

World Coal Institute. "Coal Facts, 2008 Edition." Online, December 2008. http://www.worldcoal.org/assets_cm/files/PDF/coalfacts08.pdf.

———. "Coal Meeting Global Challenges." Climate Policy Paper 1, June 17, 2009. http://www.worldcoal.org/news/coal-news/the-road-to-copenhagen/.

———. "CCS and the Post-2012 Agreement." Policy Paper 4, June 17, 2009. http://www.worldcoal.org/news/coal-news/the-road-to-copenhagen/.

———. "Key Elements of a Post-2012 Agreement on Climate Change." 2007. http://www.worldcoal.org/bin/pdf/original_pdf_file/key_elements_of_post_2012_agreement (03_06_2009).pdf.

———. "Investing in CSS." Climate Policy Paper 5, June 17, 2009(a). http://www.worldcoal.org/news/coal-news/the-road-to-copenhagen/.

Yeager, Kurt. "Congress, Think Small: We Don't Need a National Supergrid, We Need Microgrids." *AlterNet*, June 25, 2009. http://www.alternet.org/story/140911.

Abbreviations

AARP	American Association of Retired Persons
AC	alternating current
ACEEE	American Council for an Energy-Efficient Economy
AE	Austin Energy
AEO	Annual Energy Outlook
AEP	American Electric Power Company
AMI	advanced metering infrastructure
ARRA	American Recovery and Reinvestment Act of 2009
AWEA	American Wind Energy Association
B2B	business to business
BLM	Bureau of Land Management
BOS	balance-of-system
CARP	cost allocation and regional planning
CC	combined cycle
CCGT	combined cycle gas turbine plant
CCS	carbon capture and sequestration
CES	community energy system
CHP	combined heat-and-power
CIA	Central Intelligence Agency
CIM	common information mode
CO_2	carbon dioxide

CPP	critical peak pricing
CPUC	California Public Utility Commission
CSP	concentrating solar power
DC	direct current
DG	distributed generation
DNP3	distributed network protocol
DOD	U.S. Department of Defense
DOE	U.S. Department of Energy
DR	demand response
DSM	demand side management
EDF	Environmental Defense Fund
EE	energy efficiency
EERE	U.S. Department of Energy's Office of Energy Efficiency and Renewable Energy
EERS	Energy Efficiency Resource Standard
EIA	Energy Information Administration
EPA	Environmental Protection Agency
EPRI	Electric Power Research Institute
ESCO	energy service company
ESU	Energy Service Utility
FERC	Federal Energy Regulatory Commission
FPL	Florida Power and Light
GDP	gross domestic product
GHG	greenhouse gas
IBM	International Business Machines
ICCP	Inter-Control Center Communications Protocol
IEEE	Institute for Electrical and Electronics Engineers
IGCC	integrated gasification combined cycle technology
IHD	in-home display
IOU	investor owned utility
IPCC	Intergovernmental Panel on Climate Change
IRP	integrated resource planning
ISO	independent system operator
IT	information technology
ITC	investment tax credit
JCSP	joint coordinated system plan
kWh	kilowatt-hour
LEDs	light-emitting diodes
LMPs	locational marginal prices
MISO	Midwest independent system operator
MIT	Massachusetts Institute of Technology

MWh	megawatt-hour
NARUC	National Association of Regulatory Utility Commissioners
NERC	North American Electric Reliability Corporation
NESCOE	New England States Committee on Electricity
NIST	National Institute of Standards and Technology
NRDC	Natural Resources Defense Council
NREL	National Renewable Energy Laboratory
NU	Northeast Utility
NYMEX	New York Mercantile Exchange
NYSE	New York Stock Exchange
OPEC	Organization of Petroleum Exporting Countries
PBR	performance-based regulation
PG&E	Pacific Gas and Electric
PHEVs	plug-in hybrid-electric vehicles
PJM	Pennsylvania–New Jersey–Maryland Regional Transmission Organization
PNNL	Pacific Northwest National Laboratory
POLR	provider of last resort
PSCs	public service commissions
PV	photovoltaic
R&D	research and development
RECs	renewable energy credits
RES	renewable energy standards*
RNP	Renewable Northwest Project
RPS	renewable portfolio standards
RTO	regional transmission organization
SAAS	software-as-service
SCADA	supervisory control and data acquisition
SCE	Southern California Edison
SDG&E	San Diego Gas and Electric
SEIA	Solar Energy Industries Association
T&D	transmission and distribution
TCF	trillion cubic feet
TOU	time-of-use
TVA	Tennessee Valley Authority
UMTDI	Upper Midwest Transmission Development Initiative
WAN	wide-area network
W-M	Waxman–Markey

---

*Renewable portfolio standards (RPS) are requirements for obtaining a minimum percentage of electricity supplies from renewable sources placed on electricity retailers under state laws or public utility regulations. Renewable energy standards (RES) are essentially the same requirements placed on all electricity sellers nationally by federal legislation.

Acknowledgments

I AM INDEBTED to a vast number of friends, colleagues, and generous strangers who have helped me think about and then write this book. The inspiration for the book came from two projects generously sponsored by Lou Jahn and David Owens at the Edison Foundation that forced me to take a long-term look at the industry. I am also grateful to *The Brattle Group*'s management, led by Paul Carpenter, Matt O'Loughlin, and especially Hannes Pfeifenberger, without whose support the book would not have been possible. The same can be said of my editor, Todd Baldwin, who is an extraordinarily insightful reader inside a great institution.

I am greatly indebted to a number of experts and industry leaders who took time out of their busy schedules to speak with me. This list includes Greg Basheda, Richard Baxter, Paul Bonavia, Jason Bordoff, Tim Brennan, Jessica Brahaney Cain, Angela Chuang, Mark Crisson, Don Von Dollen, Glenn English, Jim Fama, Garth Corey, Jeff Genzer, Rob Gramlich, Chuck Gray, Bryan Hannegan, Steve Hauser, Bob Hemphill, Revis James, Thomas Jenkin, Jim Jura, Chris Kavalec, Melanie Kenderdine, Tom King, Karl Lewis, Richard Lynch, Che McFarlan, Richard Meserve, Vince Minni, Ernie Moniz, Mike Oldak, Karen

Palmer, Rob Pratt, Snuller Price, Rhone Resch, Jim Rogers, Joe Romm, Jeff Ross, Ted Schultz, Glen Sharp, Mike Siminovich, Rob Skidmore, Wally Tillman, Steve Troese, Fong Wan, and Lisa Wood. I also thank Drew Bittner, Anne Hampson, DeDe Hapner, Mark Johnson, Bonnie Jungerberg, Greg Kats, Sue Kelly, Troy Larson, Marty Lobel, Nicole Lynch, Lina Matsumura, Diana Moss, Ren Orans, Lawrence Pacheco, Marcia Rackstraw, Ben Rogers, Lynn Rzonca, Tami Sandberg, Omar Siddiqui, Rob Stavins, Jack Stirzaker, and Barbara Tynan.

My partners at *Brattle* were incredibly generous in their supply of references and comments, especially cherished friend Joe Wharton, Ahmad Faruqui, Phil Hanser, and Ryan Hledick (DR and the Smart Grid); Romkaew Broehm, Adam Schumacher, and Gary Taylor (my partners in California litigation); Mariko Geronimo and Dean Murphy (PHEV data); José García, Jamie Hagerbaumer, and David Robinson (integration data); and Metin Celebi, Judy Chang, Marc Chupka, and Patrick Fleming (carbon and supply side issues, notably Appendix B).

The research assistants and corporate services team at *Brattle* that helped me out, led by Laura Burns, were astonishing in their speed and capability. I especially thank Theirrien Clark, Chris Coakley, and Jenna Curto, as well as Lucas Bressan, Ahmed Sharif, and alumni Scott Hennessey, Christina Leaton, Matt McCaffree, and John Tsoukalis. While thanking all these colleagues, I emphasize that all opinions in this book are mine alone, and not necessarily those of *The Brattle Group* or these partners. None of the organizations I am affiliated with sponsored or reviewed this work prior to its completion. All errors and omissions are likewise my sole responsibility. Additions, corrections, apologies, and supplemental information on this volume (including these acknowledgments) will be posted at www.smartpowerbook.com.

Every book has its special heroes and heroines; this book has two. Marianne Gray kept my office functioning well through a brutal, unpredictable period, handled most of the manuscript, created the bibliography, and somehow managed to remain sane. Heidi Bishop put up and kept up with a torrent of research tasks that shifted from hour to hour with patience and just the right gallows humor, on top of keeping up with a wide assortment of ever-changing tasks. I will always be grateful to them.

Finally, my family has once again given me encouragement and help throughout this project. My wife, Susan Vitka, did a superb job editing many chapters, in addition to her love and moral support. I thank her, my daughter Emily, my sisters, and the rest of my family with all my heart.

Peter Fox-Penner
Washington, DC—December 2009

Index

Note: Page numbers with f, t, or b denote reference to a figure, table, or box, respectively.

Island Press | Board of Directors

ALEXIS G. SANT *(Chair)*
Managing Director
Persimmon Tree Capital

KATIE DOLAN *(Vice-Chair)*
Executive Director
The Nature Conservancy
   of Eastern NY

HENRY REATH *(Treasurer)*
Nesbit-Reath Consulting

CAROLYN PEACHEY *(Secretary)*
President
Campbell, Peachey & Associates

DECKER ANSTROM
Board of Directors
Comcast Corporation

STEPHEN BADGER
Board Member
Mars, Inc.

KATIE DOLAN
Eastern New York
   Chapter Director
The Nature Conservancy

MERLOYD LUDINGTON LAWRENCE
Merloyd Lawrence, Inc.
   and Perseus Books

WILLIAM H. MEADOWS
President
The Wilderness Society

PAMELA B. MURPHY

DRUMMOND PIKE
President
The Tides Foundation

CHARLES C. SAVITT
President
Island Press

SUSAN E. SECHLER

VICTOR M. SHER, ESQ.
Principal
Sher Leff LLP

PETER R. STEIN
General Partner
LTC Conservation Advisory
   Services
The Lyme Timber Company

DIANA WALL, PH.D.
Director, School of Global
Environmental Sustainability
   and Professor of Biology
Colorado State University

WREN WIRTH
President
Winslow Foundation